Topics in Physical Chemistry

vol 3

Edited by H. Baumgärtel, E. U. Franck, W. Grünbein
On behalf of Deutsche Bunsen-Gesellschaft für Physikalische Chemie

Topics in
Physical Chemistry

---

vol 1   Introduction to
        Surface Physical Chemistry

        K. Christmann

vol 2   Gaseous Molecular Ions
        An Introduction to Elementary Processes
        Induced by Ionization

        E. Illenberger, J. Momigny

vol 3   Liquid Crystals
        H. Stegemeyer, Guest Editor

H. Stegemeyer, Guest Ed.

Liquid Crystals

 Springer-Verlag Berlin Heidelberg GmbH

Guest Ed. address:
Prof. Dr. Horst Stegemeyer
Universität Gesamthochschule Paderborn
Institut für Physikalische Chemie
33095 Paderborn

Edited by:
Deutsche Bunsen-Gesellschaft
für Physikalische Chemie e.V.
General Secretary Dr. Heinz Behret
Carl-Bosch-Haus
Varrentrappstraße 40/42
60486 Frankfurt 90

Die Deutsche Bibliothek – CIP-Einheitsaufnahme

Liquid crystals / H. Stegemeyer, guest ed. - Darmstadt:
Steinkopff; New York: Springer, 1994
  (Topics in physical chemistry; Vol. 3)
  ISBN 978-3-662-08395-6   ISBN 978-3-662-08393-2 (eBook)
  DOI 10.1007/978-3-662-08393-2
NE: Stegemeyer, Horst (Hrsg.); GT

This work is subject to copyright. All rights are reserved, whether the whole or part of the material is concerned, specifically the rights of translation, reprinting, re-use of illustrations, recitation, broadcasting, reproduction on microfilms or in other ways, and storage in data banks. Duplication of this publication or parts thereof is only permitted under the provisions of the German Copyright Law of September 9, 1965, in its version of June 24, 1985, and a copyright fee must always be paid. Violations fall under the prosecution act of the German Copyright Law.

Copyright © 1994 by Springer-Verlag Berlin Heidelberg
Originally published by Dr. Dietrich Steinkopff Verlag GmbH & Co. KG, Darmstadt in 1994
Softcover reprint of the hardcover 1st edition 1994

Chemistry Editor: Dr. Maria Magdalene Nabbe – English Editor: James C. Willis
Production: Heinz J. Schäfer

The use of registered names, trademarks, etc. in this publication does not imply, even in the absence of a specific statement, that such names are exempt from the relevant protective laws and regulations and therefore free for general use.

Printed on acid-free paper

Preface

Since the discovery of liquid crystals in 1888 by Friedrich Reinitzer and the evidence of their existence as thermodynamically stable phases by Otto Lehmann at the beginning of this century basic research on the chemistry and physics of these mesophases has been carried out for three decades mainly in Germany and France. In the 1940s, 50s the interest in liquid crystal research, however, relaxed somewhat. But this situation abruptly changed in the middle of the 1960s when G.H. Heilmeier gave the first indication for an application of liquid crystals in electrooptical display technology. This aspect immensely stimulated liquid crystal research and led to an explosive growth of publications in this field over the last 25 years. By 1959, about 1400 compounds that form liquid crystalline phases were known; by 1992 this number had increased to about 50 000. In portable devices like wristwatches, pocket calculators, measuring instruments, and laptop computers the liquid crystal display technology gained total acceptance and is on the way to engross the market of color TV screens.

This development justifies to publish a volume devoted to liquid crystals in the series "Topics in Physical Chemistry". According to the purpose of this series, this volume must fill the gap between the scarce information about liquid crystals in physicochemical textbooks and more detailed monographs and numerous review articles written for specialists. Consequently, an easy introduction into the matter of liquid crystals should be given which will enable scientists who are not specially engaged in this field, as well as students in chemistry and physics to understand the basic problems of liquid crystals research and application. Because of the widespread field of different research activities in liquid crystals this aim was seemingly impossible for a single author trying to cover all aspects of this topic. Thus, different competent authors have been involved in writing the six chapters of this book.

In *Chapter 1* an introduction is given to the "Phase types, structures, and chemistry of liquid crystals." The position of mesophases at the temperature scale between solid crystals and isotropic liquid is emphasized together with the chemical structure of the mesogenic building blocks, both calamitic and discotic. Besides nematic and cholesteric phases which only exhibit an orientational long-range order, the variety of smectic liquid crystals exhibiting additional different types of positional long-range order is reviewed. A brief introduction into the molecular statistical theories of liquid crystals is also given.

Chapter 2 covers the "Thermodynamical behavior and physical properties of thermotropic liquid crystals." In the first part of this chapter phase transitions between the various polymorphic types of liquid crystals including the effect of pressure and critical phenomena are discussed. The following sections cover the optical, magnetic, dielectric, and elastic properties of liquid crystals. The reader will be made familiar with ferroelectricity and the flexoelectric polarization of special types of mesophases.

Chapter 3 considers "Liquid crystalline polymers." Since the discovery in the mid-1970s of liquid crystalline properties of polymers bearing mesogenic groups in their main chain and the first synthesis of liquid crystalline side chain polymers the number of papers in this field explosively increased. Calamitic as well as discotic mesogenic units are incorporated as building blocks in main chain and side chain liquid crystalline polymers, the molecular architecture of which is discussed in this chapter together with the phase behavior of the resulting nematic, cholesteric, and smectic phases. The following sections cover the physical properties of liquid crystal polymers (optical behavior, mechanics, rheology, dynamic behavior, and electric field effects) and the theories describing these phases.

Chapter 4 addresses "Lyotropic liquid crystals." The first observation of lyotropic liquid crystals dates from 1854 when Virchow described the behavior of myelin forms, however, without any knowledge about the physical nature of this state of matter. Generally, lyotropic liquid crystals are formed from amphiphilic compounds dissolved in a solvent (mostly water) which organize in anisometric aggregates (micelles). At higher solute concentrations different liquid crystalline phases exist (nematic, cubic, hexagonal, lamellar). The structure, anisotropic properties, and phase diagrams are discussed in this chapter together with theoretical aspects of the aggregation of amphiphilics.

Chapter 5 deals with "Application of liquid crystals in spectroscopy." Liquid crystals, especially nematics, are solvents well suited to the study of the anisotropic properties of solute molecules which are oriented within the anisotropic matrix. The application of this method in UV, IR, NMR, ESR, Mössbauer and fluorescence spectroscopy is discussed, including sample preparation technique and the determination of order parameters.

In *Chapter 6* an overview is given of the "Application of liquid crystals in display technology." The worldwide breakthrough of liquid crystals in display technology is based on two facts: the extremely small power consumption and the very low control voltage. The electrooptical field effects for nematic liquid crystal displays are illustrated in this chapter together with the driving operation technique (time multiplexing, actively addressed matrix technique). Modern developments like supertwisted nematic displays for high information content and fast response are discussed. Field effects in chiral ferroelectric smectics are reviewed; they allow very short switching times but are still far from presenting a technological breakthrough. Recently, liquid crystal display projection systems opened new aspects in large scale information systems.

Finally, I hope this volume will allow an introduction to the field of liquid crystals for all scientists and students who want to delve into this "fourth state of matter." I wish to thank all the authors who contributed to complete this book, especially Prof. G. Pelzl for his extensive revision of Chapter 1, and the chemistry editor of the publisher, Dr. Maria Magdalene Nabbe, for an effective collaboration and for her endurance during the preparation of this volume.

Horst Stegemeyer   Paderborn, December 1993

Contents

Preface V

1	**Phase Types, Structures, and Chemistry of Liquid Crystals**	1
	D. Demus	
1.1	Historical	1
1.2	The Position of the Liquid Crystals between the Solid and the Isotropic Liquid State	2
1.3	Types of Liquid Crystals	4
1.4	Thermotropic Liquid Crystals	6
1.4.1	Calamitic Liquid Crystals	6
1.4.1.1	Phase Structures and Textures of Calamitic Liquid Crystals	6
1.4.1.2	Sequence Rule and Reentrant Behavior	26
1.4.1.3	Chemistry of Calamitic Liquid Crystals	26
1.4.1.4	Molecular Statistical Theories of the Calamitic Phases	35
1.4.2	Discotic Liquid Crystals	42
1.4.2.1	Phase Structures of Discotic Liquid Crystals	42
1.4.2.2	Chemistry of Discotic Liquid Crystals	44
1.4.2.3	Molecular Statistical Theory of Discotic Phases	46
1.4.3	Sanidic Liquid Crystals	47
1.4.4	Cubic Mesophases	49
References		50
2	**Thermodynamic Behavior and Physical Properties of Thermotropic Liquid Crystals**	51
	G. Pelzl	
	Introduction	51
2.1	Calorimetric Investigations	51
2.2	Optical Anisotropy	57
2.2.1	Optically Uniaxial Liquid Crystals	57
2.2.2	Optically Biaxial Liquid Crystals	63
2.3	Optical Properties of Twisted Liquid Crystalline Phases	65
2.3.1	Cholesteric Phases	65
2.3.2	Chiral Smectic C Phases (S_C^*)	67
2.4	Magnetic Properties	68
2.5	Dielectric Properties	70
2.5.1	Static and Low-Frequency Dielectric Anisotropy	70
2.5.2	Frequency Dependence of the Dielectric Permittivity	73
2.6	Elastic Properties of Liquid Crystals	77
2.6.1	Elastic Properties of Nematic Liquid Crystals	77
2.6.2	Uniform Alignment of Nematic Liquid Crystals by Wall Forces	78
2.6.3	Deformation of Nematic Liquid Crystals by an Electric Field	80

2.6.4	Deformation of Cholesteric Liquid Crystals by Electric Fields	84
2.6.5	Deformation of Smectic Liquid Crystals (S_A, S_C) by an Electric Field	84
2.7	Ferroelectric Properties	88
2.8	Flexoelectric Polarization	90
2.9	Diffusion Coefficients	91
2.10	Electric Conductivity	92
2.11	Electrohydrodynamic Instabilities	95
2.12	Viscosity	99
References		101
3.	**Liquid Crystalline Polymers**	103
	R. Zentel	
	Introduction	103
3.1	Liquid Crystalline Main Chain Polymers	104
3.1.1	Thermotropic Main Chain LC-Polymers	104
3.1.1.1	Thermotropic LC-Polyesters	105
3.1.1.2	Semiflexible Main Chain LC-Polymers	107
3.1.2	Lyotropic Main Chain LC-Polymers	108
3.1.2.1	Theory	110
3.1.2.2	Rheology	113
3.1.2.3	Application	114
3.2	Liquid Crystalline Side Chain Polymers with Calamitic Mesogens	116
3.2.1	Structures and Structure-Property Relations	116
3.2.2	Interaction of Polymer Chains and Mesogens	120
3.2.3	Influence of Additional Structural Variations on the Properties	123
3.2.4	Physical Properties of the Liquid Crystalline Phases	127
3.3	Liquid Crystalline Polymers with Disc-like Mesogens	133
3.4	Liquid Crystalline Elastomers	136
References		139
4	**Lyotropic Liquid Crystals**	143
	K. Hiltrop	
4.1	Introduction and Historical Background	143
4.2	Hydrophobic Effect and Aggregation	145
4.2.1	The Hydrophobic Effect	145
4.2.2	The Aggregation	146
4.3	Liquid Crystalline Structures and Nomenclature	150
4.4	Phase Diagrams	153
4.5	Physical Properties and Investigation Methods	156
4.5.1	Polarizing Microscopy	156
4.5.2	Nuclear Magnetic Resonance	158
4.5.3	Small-Angle Scattering of X-rays (SAXS) and of Neutrons (SANS)	160
4.6	Theoretical Aspects	162

4.6.1	Interactions	162
4.6.2	The R-theory	164
4.6.3	Packing Considerations	164
4.6.4	Other Theories	166
4.7	Applications	168
References		169
5	**Application of Liquid Crystals in Spectroscopy**	173
	L. Pohl	
	Introduction	173
5.1	Molecular Orientation	173
5.2	Selection of Appropriate Solvents	175
5.3	Preparation of Nematic Solutions and of Samples	175
5.4	Infrared Spectroscopy in Liquid Crystals	177
5.5	Determination of the Order Parameter and of the Polarization	179
5.6	Visible and Ultraviolet Spectroscopy in Liquid Crystals	180
5.7	Fluorescence Spectroscopy in Liquid Crystals	182
5.8	Nuclear Magnetic Resonance Spectroscopy in Liquid Crystals	184
5.9	Electron Spin Resonance Spectroscopy in Liquid Crystals	188
5.10	Mössbauer Spectroscopy in Liquid Crystals	191
References		193
6	**Liquid Crystal Displays**	195
	M. Schadt	
	Introduction	195
6.1	Electrooptical Field Effects for Nematic LCDs	197
6.1.1	Liquid Crystal Field Effects in General	197
6.1.2	Wave-Guiding Operation of Twisted Nematic Liquid Crystal Displays (TN-LCDs)	200
6.1.3	Time Multiplexing of LCDs	204
6.1.4	Operation of TN-LCDs with Circularly Polarized Light	206
6.1.5	Supertwisted Nematic LCDs and LC Materials	206
6.1.6	The Color Problem of Supertwisted LCDs	210
6.1.7	Full Color, Actively Addressed TN-LCDs for High Information Content and Fast Response	213
6.2	Field Effects in Chiral Ferroelectric S_C^* Liquid Crystals	215
6.3	Recent Developments in LCD Projection Systems	219
	Conclusion	223
References		224
Subject Index		227

Authors' addresses

Prof. Dr. Dietrich Demus
Chisso Petrochemical Corporation
Tchikara-shi
Chibakon 290/Japan

Dr. Karl Hiltrop
Universität Gesamthochschule
Paderborn
Institut für Physikalische Chemie
33095 Paderborn

Prof. Dr. G. Pelzl
Martin-Luther-Universität
Halle Wittenberg
Institut für Physikalische Chemie
Mühlpforte 1
06108 Halle

Dr. Ludwig Pohl
E. Merck
64271 Darmstadt

Dr. Martin Schadt
F. Hoffmann-La Roche AG
CH-4002 Basel
Schweiz

Prof. Dr. Rudolf Zentel
Johannes-Gutenberg-Universität Mainz
Institut für Organische Chemie
Johann-Joachim Becher Weg 18-22
55128 Mainz

1 Phase Types, Structures and Chemistry of Liquid Crystals

D. Demus

1.1 Historical

Liquid crystals are well known today to a broad community. Liquid crystal displays in electronic watches and calculators are wide spread. However, much less is commonly known about the function of these displays or the scientific background of the liquid crystalline state which may be considered as a special physical condition.

The liquid crystalline state was detected more than 100 years ago. In 1888, Friedrich Reinitzer, an Austrian botanist and chemist at the University of Graz, synthesized several esters of cholesterol, a natural product occurring in plants and animals. Reinitzer found in these esters the phenomenon of "double melting," i.e., at a certain temperature the compound changes from the crystalline solid phase to an opaque liquid which transforms at a defined higher temperature to an optically clear liquid. These phase changes were reproducible with increasing and decreasing temperature in several compounds. Looking to the literature before Reinitzer, some indication can be found that several scientists indeed dealt with liquid crystals, but did not notice the unique phenomena and therefore did not become aware of this new state.

Reinitzer himself was not able to explain the curious phenomenon of "double melting" and the existence of the opaque liquid. Therefore, he sent samples of his compounds to Otto Lehmann, who was professor of physics (successor of Heinrich Hertz) at the Technical High School of Karlsruhe. Lehmann, at that time, was the leading crystallographer in Germany, and quickly found the optical anisotropy of the opaque liquid phases of Reinitzer's cholesterol esters. Despite the fact that the chemical constitution of cholesterol was elucidated much later, Lehmann intuitively argued that the optical anisotropy of these liquids would be due to elongated molecules which are oriented parallel with the long axes. Today, it is clear that, in principle, this explanation was valid. Lehmann created the designation "fluid crystals" and "liquid crystals" ("fließende Kristalle," "flüssige Kristalle").

Compounds which exhibit the liquid crystalline state in a certain temperature interval are called "thermotropic" liquid crystalline. Lehmann and others also experimented with mixtures, especially mixtures of certain organic salts and water, and detected a variety of liquid crystals which is existent only in mixtures of this kind and which are called "lyotropic" liquid crystals.

Over the next decade about 15 liquid crystalline compounds became known; in all these cases the detection of the liquid crystalline phases occurred by chance since no connection between the molecular shape and the liquid crystalline state was known.

Soon after 1900, Daniel Vorländer, who was professor of chemistry at the University of Halle, started his systematic synthetic work in order to find connections between the molecular structure of chemical compounds and the occurrence of the liquid crystalline state. Proved by many examples and counterexamples, Vorländer, already in 1908, was able to establish his rule that liquid crystalline compounds must have a molecular shape as linear as possible. Until Vorländer's retirement in 1935, in his laboratory about 1100 liquid crystalline substances were synthesized; they constituted about 90% of the liquid crystals known until 1960. Vorländer may be considered the "father of the liquid crystal chemistry". In 1906, he had detected the phenomenon of liquid crystalline polymorphism, i.e., that a given compound exhibits more than one liquid crystalline phase.

In the first three decades of this century, especially French crystallographers were very active in liquid crystal research. The most outstanding contribution to their work was by G. Friedel (University of Strasbourg), who gave the first rational explanation for the pictures observed in the polarizing microscope, and concluded from his observations on the structure of these phases (and this was before application of x-ray structure investigations on the liquid crystalline state).

Systematic synthetic work and physical investigations of many different properties of the liquid crystals were done up to 1965. Important advances included the theoretical explanation of this state based on a model of long stiff rods by Onsager (1949) which pointed at the dominant role of the repulsive forces, and a molecular statistical consideration by Maier and Saupe (1960) which showed the importance of the dispersion forces. Both aspects have been united in the van der Waals theories, which have been elaborated on in the last 15 years, especially by Gelbart and Cotter.

Induced by the applications of liquid crystals in optoelectronic displays and for thermography in the mid-1960s, liquid crystal research increased exponentially and spread from a few centers to many institutions in all developed industrial countries. Due to strong success in the practical as well as the theoretical investigation of the liquid crystalline state in the 1970s there seemed to be a certain stagnation in "spectacular" findings until 1977, when Chandrasekhar was able to show that not only rod-like molecules, but also compounds with disc-like molecular shape are able to form liquid crystals of different phase structures. In recent years the gap between rod-like and disc-like molecules could be filled with lath-like molecules, which also are able to form liquid crystalline phases.

1.2 The Position of the Liquid Crystals between the Solid and the Isotropic Liquid State

Molecular crystals possess a long-range order of the positions of the molecules and, in addition, a long-range order of their orientations. With increasing temperature – at the melting point – the crystal transforms to the isotropic liquid. This may be expressed by a transition scheme:

crystalline solid ↔ isotropic liquid

The respective phase transition is of first order and, in principle, is reversible, i.e., with decreasing temperature from the isotropic liquid the solid phase can again be obtained by the crystallization process.

Liquid Crystals

The designation "liquid crystal" has been enormously extended in recent years, which is partly due to the detection of novel molecular types of liquid crystalline substances, and partly due to the discovery of additional phase structures.

Compared to isotropic liquids, liquid crystals represent a higher state of order which may be expressed in terms of order parameters. Compared to solid crystals, the liquid crystals have a higher intermolecular and intramolecular mobility, i.e., they may have several degrees of freedom of molecular rotation, translation, oscillation, and intramolecular confirmational changes. As mentioned above, some of these mobilities are also allowed in solids. Therefore, the exact delimitation between liquid crystals and solids is not clear.

Cubic Plastic Crystals

If the molecules of the substance have more or less spherical shape there may be an additional phase transition in the solid state, leading to a "plastic phase". This plastic phase is three-dimensional with respect to the positional order, however, due to rotational freedom of the molecules there is diminished orientational order. Well known examples for this behavior are methane, carbon tetrachloride, campher or especially carbon tetra bromide CBr_4 with the following transition scheme:

crystalline solid (rhombic) 47° C plastic crystal (cubic) 94° C isotropic liquid

Since these plastic phases have cubic symmetry they are optically isotropic. Their viscosity and mechanical stability (elastic constants) are several orders of magnitude lower than those of ordinary crystalline solid phases.

Hexagonal Plastic Crystals

There are compounds with elongated molecular structure in which, at certain temperatures, a "rotational transition" takes place and the whole molecules are allowed to rotate around the molecular long axis. By this diminished state of order the symmetry of the phase may become higher and in many cases hexagonal "plastic" crystals exist.

As long known prominent examples for this behavior, the higher n-alkanes may be mentioned, e.g.,

$n\text{-}C_{23}H_{48}$

crystalline solid 41°C plastic crystal 50°C isotropic liquid

Especially in the case of the n-alkanes, it is an open question if the plastic phases of these materials are really solids or if they can be attached to one of the highly ordered liquid crystalline phases. This special problem leads to a general question: What is the borderline between the solid crystals and the liquid crystals. There is no unique answer to this question. Depending on the method of consideration certain structures with three-dimensional order may be considered as solids or as highly ordered liquid crystals (see Section 1.4.1.1).

1.3 Types of Liquid Crystals

Liquid crystals are also called crystalline liquids, mesophases or mesomorphic phases. Compounds with mesomorphic properties may be called "mesogenic".

General Types
There are *thermotropic liquid crystals*, in which the liquid crystalline phases exist in certain temperature regions, e.g., the famous 4,4′-azoxyanisole:

$$CH_3O-\bigcirc-\underset{O}{N=N}-\bigcirc-OCH_3$$

crystalline solid 118°C liquid crystalline 135°C isotropic liquid.

The temperature at which the liquid crystalline state is transformed into the isotropic liquid is generally designated as clearing temperature. The liquid crystalline phase is a nematic (N) one (see Section 4.1.1.1). The transition schemes may be given in an abbreviated form with the transition temperatures in °C:

cr 118 N 135 is,

which will be used also in the following.
Thermotropic liquid crystalline phases may occur in pure compounds, but also in mixtures.
There is another general type of liquid crystal which only exists in mixtures consisting of compounds with a relatively high polarity (ampiphilic compounds) and certain solvents. They are called *lyotropic liquid crystals* and the necessary condition for their existence is a strong interaction of the polar compounds with the molecules of the solvent. Well known examples are mixtures of alkali n-alkanoates (soaps) and water.
Certain compounds are able to form thermotropic as well as lyotropic liquid crystals, they are called "amphotropic".
Alkali alkanoates are examples for amphotropic behavior.

Molecular Structure
Derived from the shape of the constituting molecules, we may differentiate among:
calamitic liquid crystals occurring in rod-like molecules, which may be considered, as the classical liquid crystals (Section 1.4.1),
discotic liquid crystals derived from disc-like molecules, which were discovered in 1977 (Section 1.4.2),
sanidic liquid crystals occurring in lath-like (board-like) molecules, which were first found in 1986 (Section 1.4.3).

From the standpoint of the molecular size the differentiation in low molecular (monomeric) and polymeric mesomorphic compounds (see Chapter 3) is possible, e.g.,

monomer

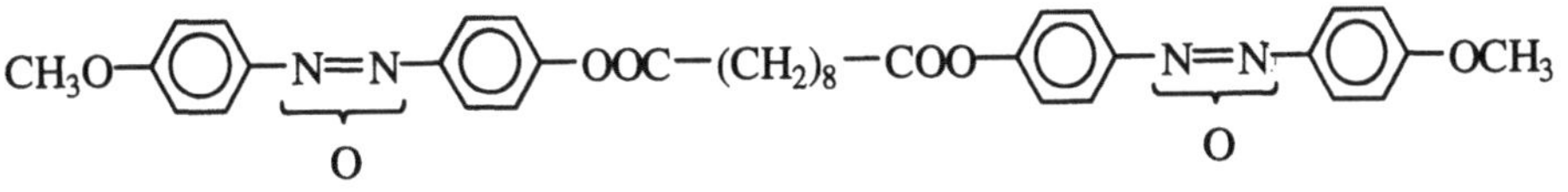

$$CH_3O-\bigcirc-N=N-\bigcirc-OOC-(CH_2)_8-COO-\bigcirc-N=N-\bigcirc-OCH_3$$
$$OO$$

crystalline solid 149°C liquid crystalline 221°C isotropic liquid

polymer

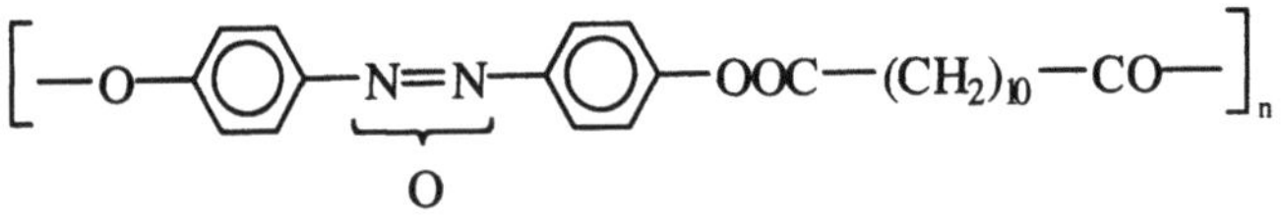

$$\left[-O-\bigcirc-N=N-\bigcirc-OOC-(CH_2)_{10}-CO-\right]_n$$
$$O$$

crystalline solid 216°C liquid crystalline 265°C isotropic liquid

Phase Structure
The molecular structure has a strong influence on the phase structure of the liquid crystals. On the other hand, similar phase structures may occur in compounds of very different shape, and in one and the same compounds many different phase structures may be realized.
The most important phase structures of liquid crystals are:
nematic, the most liquid-like structure in which, contrary to isotropic liquids, one or two molecular axes are oriented parallel to one another, resulting in an orientational long-range order;
smectic, layer structures with many possibilities of the state of order inside the layers and different possibilities of mutual arrangement of the layers, showing long range orientational and more or less positional order;
cubic, structures with micellar lattice units or complicated interwoven networks;
columnar, structures with columns consisting of parallel arranged disc-like molecules.

1.4. Thermotropic Liquid Crystals

1.4.1. Calamitic Liquid Crystals

1.4.1.1 Phase Structures and Textures of Calamitic Liquid Crystals

Nematic Liquid Crystals
At present about 20 000 compounds with nematic phases are known. The word nematic (Greek $\nu\eta\mu\alpha$ = thread) refers to thread-like defects often observed in micrographs. A typical example is 4-methoxybenzylidene-4'-n-butylaniline (MBBA):

CH$_3$O—⟨◯⟩—CH=N—⟨◯⟩—C$_4$H$_9$ cr 22 N 47 is

The molecules of this compound are elongated and linear (see Fig. 1.1). Since, in the nematic state, the molecules have rotational freedom around the long molecular axis, the effective molecular shape is not flat (despite the flat character of the benzene rings), but may be better described by a rotational cylinder. Therefore, in most of the theoretical models of nematic liquid crystals the molecules are approximated by rotational cylinders. In simple schemes which represent only the main features of a liquid crystal phase structure in comparison to the isotropic liquid, even rods may be used as molecular models (see Fig. 1.2). It can be seen in this figure that, contrary to the isotropic state, the molecular long axes in the nematic state are, on average, parallel; however, the molecules may shift freely in the direction of the long axis, i.e., the nematic state is characterized by long-range orientational order and short-range positional order. The molecules can rotate about the long axes and short axes, and in many compounds the molecules have several possibilities of intramolecular mobility, and especially conformational changes are allowed. With respect to the intramolecular mobility the molecules in nematic phases show similar behavior as in isotropic liquids.

Because of the high mobility the nematic phases have low viscosities very similar to those of isotropic liquids. From the standpoint of viscosity, they are clearly liquid-like, however, because of the parallelity of the molecular long axes they exhibit anisotropy of many physical properties. Nematic liquid crystals are anisotropic with respect to the optical properties (double refraction), viscosity, electrical, and magnetical susceptibility, electrical and thermal conductivity (see Section 2.2, 2.4, 2.5, 2.10, 2.12). The parallelity of the molecules is not exact, especially single molecules may have strong deviations from it. On the other hand, the parallelity is strongly temperature dependent. The order parameter S is a quantitative measure for the parallelity of the molecules in liquid crystalline phases:

$$S = \tfrac{1}{2}\langle 3\cos^2\beta - 1\rangle.$$

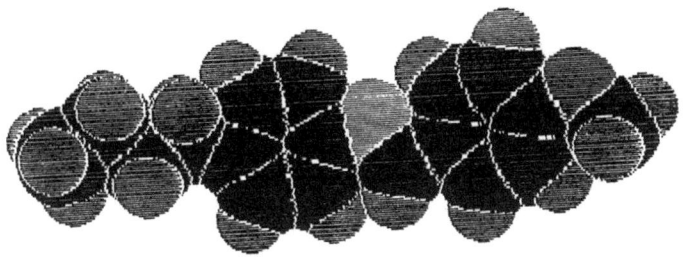

Fig. 1.1. Molecular model of 4-methoxybenzylidene-4'-n-butylaniline (MBBA).

The mean value of the direction of the molecular long axes is called the director $\vec{n}$; β is the angle between the long molecular axis and the director, brackets indicate the average.

In isotropic liquids S is zero, in structures with exact parallelity of the axes it equals one. In nematic phases S increases from values of about 0.3 near to the clearing temperature to 0.6 or even 0.7 at low temperatures (see Fig. 1.3). The order parameter can be measured by the aid of anisotropic properties, especially optical measurements (double refraction), magnetic properties, x-ray and NMR investigations.

As a result of the parallelity of the molecular long axes, nematic liquid crystals possess an intrinsic optical anisotropy (double refraction). From the optical standpoint, nematics are uniaxial with positive character, the optical axis being parallel to the director (see Section 2.2.1). The observation of liquid crystals is best made by use of a polarizing microscope equipped with a heating stage, with the liquid crystal samples in thin layers (about 10 μm) between two glass slides. Depending on the orientation of the liquid crystal, different pictures, called "textures," may be observed. Often, these textures are characteristic and may be used for identification of the liquid crystal type.

Homogeneously aligned nematics (nematic "single crystals") do not exhibit specific textures. The whole observed area is homogeneously colored which is caused by the well-known interference colors of double refracting media between crossed polarizers. By observation in direction of the optical axis the preparation will be dark (pseudoisotropic or homeotropic orientation).

With ordinary glass, i.e., without special treatment of the glass surface, characteristic textures may be obtained, e.g., the nematic Schlieren texture in Fig. 1.4. This texture is characterized by the occurrence of dark points from which two resp. four dark Schlieren begin. These dark points are the projections of special vertically oriented defect lines, called disclinations. The disclinations are nearly unknown in the case of solid crystals, however, because of the small values of the elastic constants (see Section 2.6.1) in liquid crystals they are wide spread in this state. The structure of defects may be indicated by the director distribution which is given in Fig. 1.5 for two

Fig. 1.2. a) Structure of the isotropic phase **b)** nematic phase.

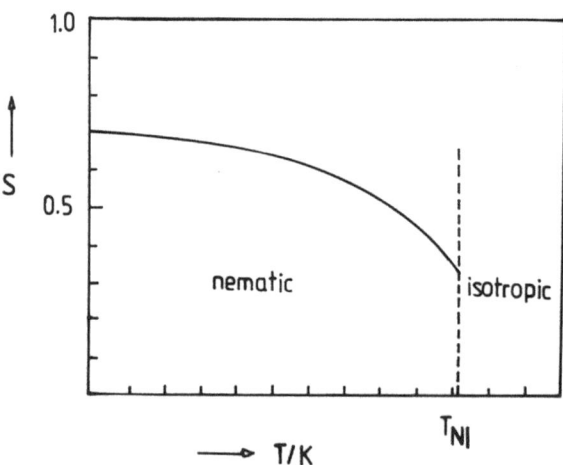

Fig. 1.3. Temperature dependence of the order parameter S in nematic phases.

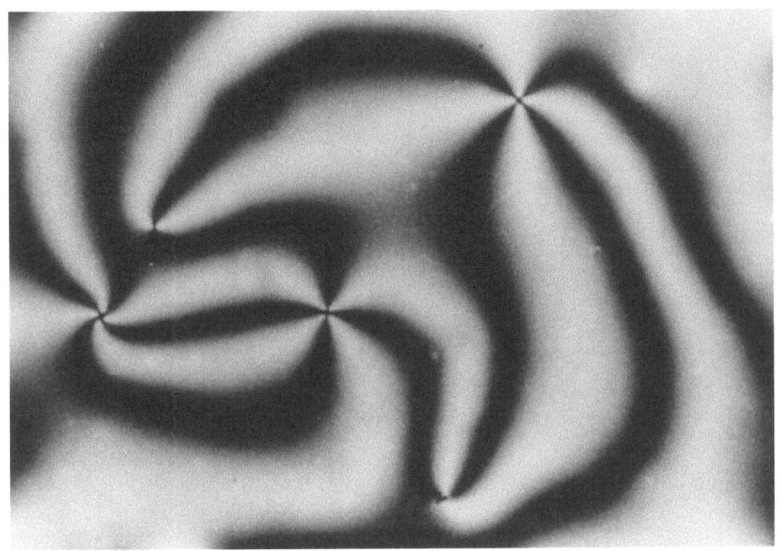

Fig. 1.4. Nematic Schlieren texture.

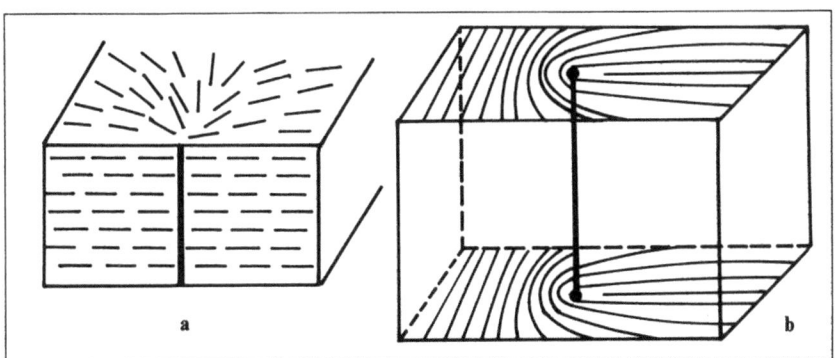

Fig. 1.5. Structure of defects in nematic phases: a) π disclination b) $\pi/2$ disclination. The orientation of the director is indicated by lines.

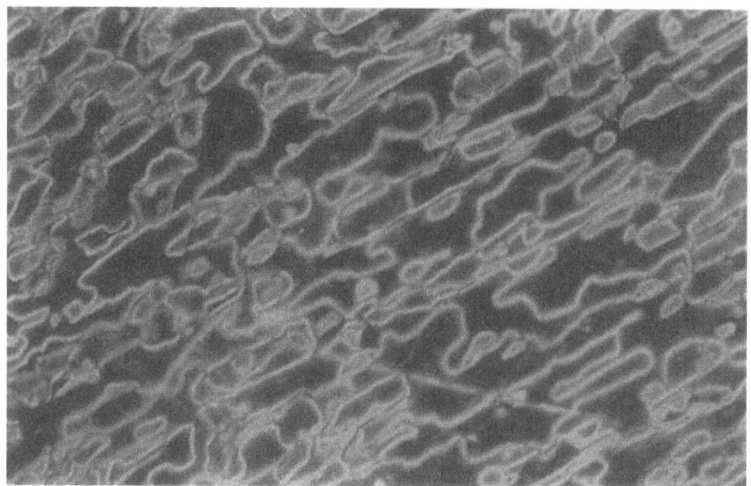

Fig. 1.6. Nematic thread-like texture.

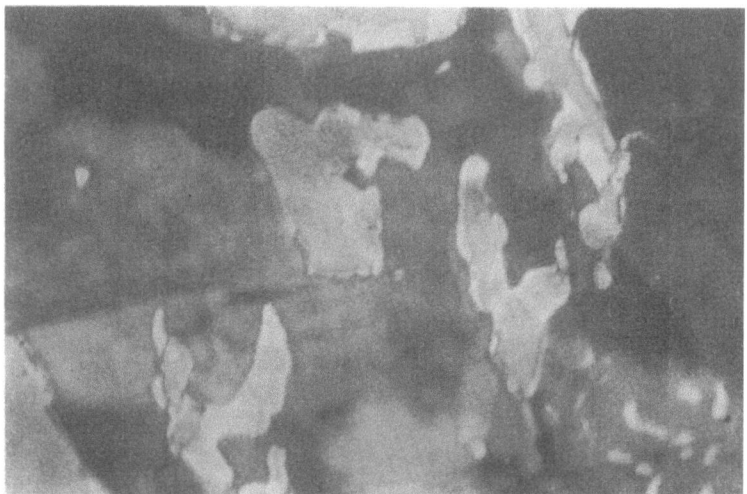

Fig. 1.7. Nematic marbled texture.

cases of nematic disclinations. The frequent existence of the defect lines and the occurrence of the dark Schlieren proves that the nematic medium in ordinary preparations usually will be highly distorted. The distortion needs additional energy, but is stabilized by interaction with walls or dust particles.

In Fig. 1.6 the "threaded" texture is shown which gave rise to the term nematic. Figure 1.7 presents the nematic "marbled texture". Its structure is mainly determined by strong interaction of the thin nematic layer with the surrounding walls. In fact, this nematic texture is a copy of the solid surface and may be used for systematic surface analyses.

Cholesteric Liquid Crystals

The cholesteric state is a variant of the nematic state occurring in chiral compounds. The designation "cholesteric" is derived from the fact that this phase type was first observed in derivatives of cholesterol. The symmetry of the molecules is reproduced in force fields of low symmetry which lead to a twisted structure, see Fig. 1.8. The directors of the neighboring quasi-nematic layers are turned at a constant angle. In this manner the medium has a twisted structure. Similar to screws, right-and left handed structures are possible. The distance of two layers turned at an angle of 360°C corresponds to the pitch p of the screw.

The strong relationship to the nematic phases is expressed by the fact that, by use of electrical or magnetical fields or surface interaction, the twisted structure may be enrolled to give ordinary nematic liquids. On the other hand, it is possible by interaction with the surrounding walls to impose a mechanical twist to nematics which yields cholesteric structures with large pitches. Nematics deformed in this manner are the basis of all practically used liquid crystal displays (see Chapter 6).

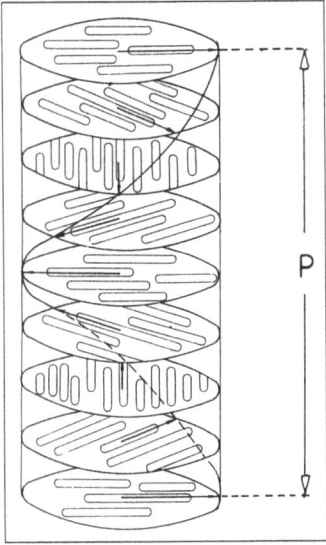

Fig. 1.8. Structure of cholesteric phases (N*); p = pitch of the helix.

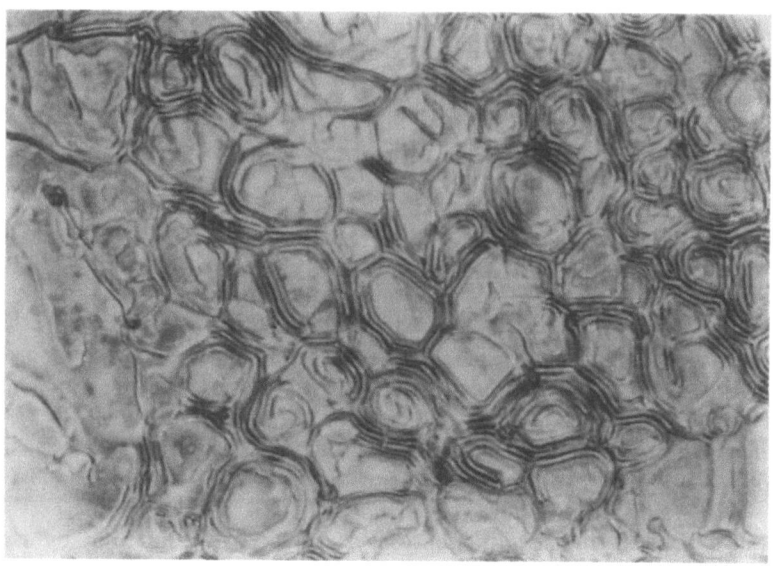

Fig. 1.9. Cholesteric texture

The twisted structure of cholesterics is the reason for several unique optical properties, namely, extremely high optical activity, selective reflection of circularly polarized light (see Section 2.3.1). The cholesterics are optically uniaxial with negative character (Section 2.2.1).

Many other properties of the cholesterics, e.g., thermodynamic properties, elastic constants are similar to the nematic state.

Because of the relatively complicated structure which may easily be deformed by even small forces, the textures of cholesterics show many peculiarities. Here, only one typical example containing the so-called "oily streaks," which are a series of defects, is given in Fig. 1.9. Cholesterics with large twist, in a small temperature interval between the isotropic state and the cholesteric phase, can exhibit up to three different "blue" phases. The blue phases show cubic symmetry, are optically isotropic, show optical activity and selective reflection of circularly polarized light. They exhibit strong light scattering which, by irradiation with white light, may cause a blue to green color, giving rise to the designation for this curious state. The structure of this state has not been completely elucidated, the most probable models explain the structure in terms of defects in a cubic array.

Smectic Liquid Crystals

Rod-like molecules are able to form liquid crystalline phases where, additional to the orientational order of the molecular long axes, the centers of gravity are, on average,

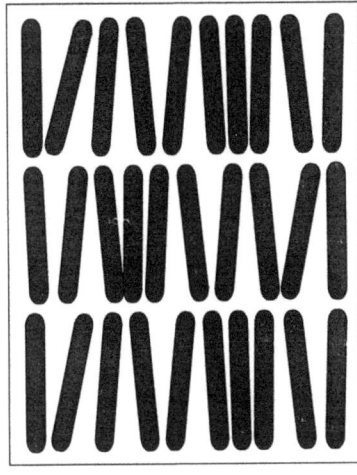

Fig. 1.10. Structure of smectic A phases.

arranged in equidistant planes so that a stratified structure results. Such phases with a stratified structure are called "*smectic*". This term comes from the Greek σμηγμα (soap), because at first such liquid-crystalline phases were observed on ammonium and alkali soaps.

There are many different types of smectic phases. According to the chronological sequence of their detection the smectic phases have been designated with code letters A, B, C, ... M.

Smectic phases without order within the layers: smectic A and C

Smectic A phases have a layer structure. Inside the layers the molecules are parallel, on an average, one to each other with their long axes perpendicular to the layer plane (Fig. 1.10). Except for the organization in layers, the molecules have a nematic-like state of order, i.e., they have the possibility for rotation around the long and short molecular axes, translational diffusion, and, within the layer, no long-range order with respect to the positions.

Basic information about the structure of the smectic A phase may be obtained from x-ray investigations. Figure 1.11 presents a sketch of x-ray diffractograms obtained in a Debye-Scherrer camera. In a non-oriented sample the local directors are turned at arbitrary angles to each other, i.e., the preparation is comparable to a polycrystalline sample. As Fig. 1.11a shows, there are two reflexes: an outer blurred halo at a scattering angle of about $\vartheta = 10°$ and an inner sharp reflection at about $\vartheta = 1.5°$. Using the equation of Bragg

$n\lambda = 2d\sin\vartheta,$

(λ = wave length of the x-ray radiation, n = order of reflection, d = translation period), it is possible to evaluate the x-ray diagram. The outer diffuse ring corresponds to a translation period of about 0.4 nm which is equivalent to the averaged

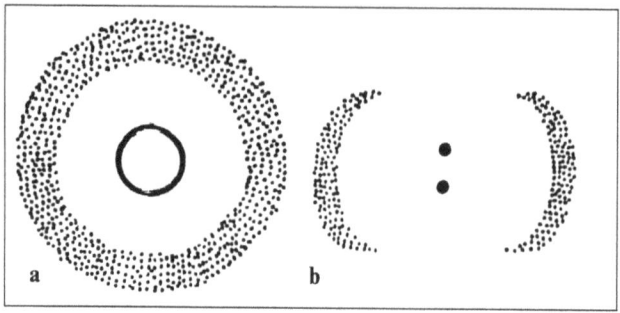

Fig. 1.11. X-ray pattern of a smectic A phase: a) non-oriented sample; b) oriented sample.

lateral distance of the molecules; the inner sharp reflection corresponds to the layer thickness with a period d which can be about the molecular length. In the case of an oriented sample – comparable to a monocrystal – the diffuse outer crescent and the sharp inner ring split up (Fig. 1.11b). The diffuse outer ring points to the fact that the parallelity of the molecules is not perfect, generally in smectic A phases the orientational order parameter S lies between 0.5 and 0.8.

In many smectic A phases the layer thickness is smaller than the molecular length. Since this effect is especially pronounced in compounds with long alkyl chains, this points to conformations of the chains which are not all trans. These "molten" alkyl chains are characteristic for all low ordered liquid crystals.

In special cases, e.g., compounds with strongly polar groups (longitudinal cyano or nitro substituents) the layer thickness may exceed the length of a single molecule. In this case the layer structure is characterized by an additional ordering which is due to an antiparallel alignment of the molecules.

Since there are additional variants of smectic A phases with undulation of the layers, there is a rich polymorphism, even in this simplest case of smectic phases. Smectic A phases are wide spread, many nematogenic compounds additionally at lower temperatures show smectic A phases.

Smectic C phases are closely related to A phases. The most important difference is the tilt of the molecular long axes with respect to the layer normal (Fig. 1.12). The mobility of the molecules is similar to that of the A phases, however, the rotation around the long axis is slightly more hindered than in A phases. In non-oriented samples the x-ray diffractogram does not show differences to that of an A phase (Fig. 1.13a), however, in oriented samples due to the tilted structure the middle lines of the two pairs of reflexes are no longer perpendicular (Fig. 1.13b), the angle between them corresponds directly to the tilt angle. Because of the tilt (angle θ) the layer thickness d cannot be equal to the molecule length L, but the relation

$$d = L \cos \theta$$

approximately holds.

Fig. 1.12. Structure of smectic C phases.

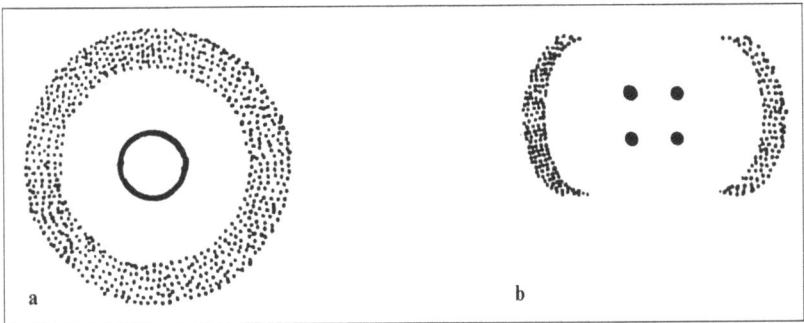

Fig. 1.13. X-ray pattern of smectic C phases: a) non-oriented sample; b) oriented sample.

The tilt angle is strongly temperature dependent and decreases to zero at the phase transition to a smectic A phase in compounds which exhibit both these phase types.

The smectic A and C phases are able to exist in several texture variants. If the A phase of a compound is homeotropic (i.e., the layers are parallel to the supporting glass plates) the texture is dark between crossed polarizers, the corresponding texture of the C phase is a Schlieren texture (Fig. 1.14). In this texture the layers are still parallel with respect to the glass plates. The Schlieren, however, indicate strong distortions of the director field.

Since smectic A phases are optically uniaxial the homeotropic texture extinguishes the light between crossed polarizers. The C phase, because of its monoclinic symmetry, is optically biaxial, therefore, it is practically impossible to have homeotropic textures.

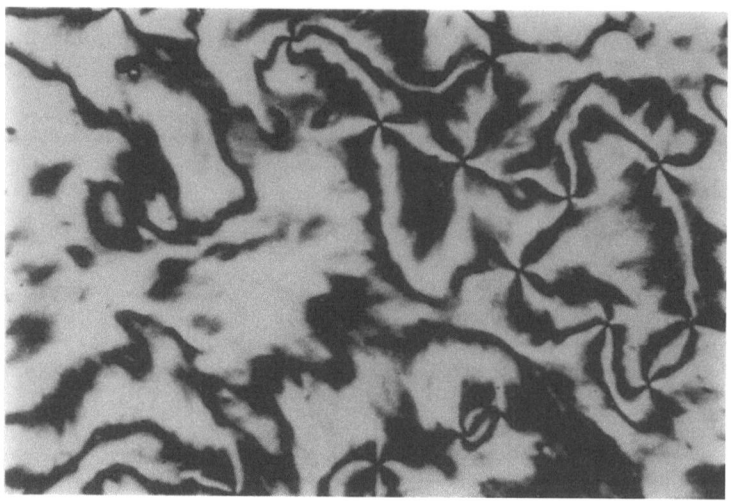

Fig. 1.14. Smectic C Schlieren texture.

In most cases the layers of the smectic A phases are not flat, but rather are distorted. As Fig. 1.15a indicates, the layers may be enrolled to give cylindrical structures. If the cylinders themselves are bent and there are many layers, the so-called anchor-ring arises (Fig. 1.15b) which has two geometrical lines of maximum distortion of the structure: a straight line and a circle, both of which are visible in the respective textures because they are places of maximum light scattering. If the anchor-ring is slightly deformed, the so-called "Dupin cyclide" is existing (Fig. 1.15c). In textures containing cyclides (called "focal conic" textures) branches of hyperbolas and ellipses exist as geometrical places of maximum deformation. Both of these, in fact, are visible in special orientations. The hyperbolas can be found in the wide spread fan-shaped texture, (Fig. 1.16) which has been known since Lehmann's time and was analyzed in much detail by the French crystallographer G. Friedel. In another orientation which preferably exists in somewhat thicker preparations, the polygonal texture with clearly developed ellipses is obtained (Fig. 1.17).

Because of the close relation between smectic A and C phases the latter also are able to form focal conic textures, however, due to the tilted structure of the C phases their textures have additional discontinuities. Figure 1.18 presents the "broken fan-shaped" texture. The split of the domains in a conglomerate of many small regions of homogeneous color is due to the different direction of the tilt in these regions.

Smectic phases with order within the layers: smectic B, F, G, ... M

All of the smectic phases explained below possess layer structures; contrary to the smectic A and C phases the molecules show some positional order within the layers, depending on the phase type.

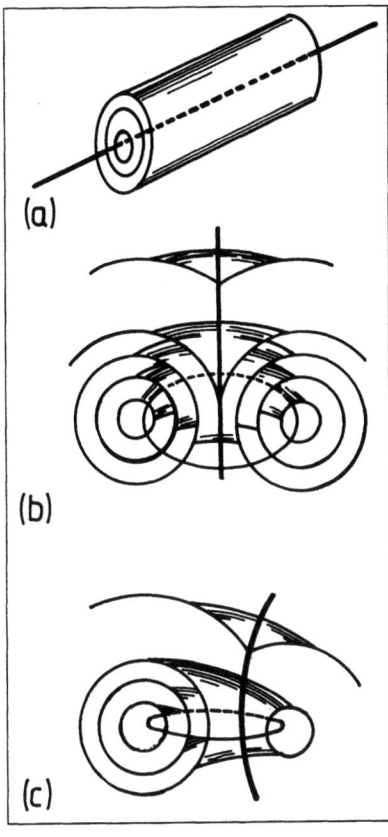

Fig. 1.15. Defects in smectic A phases: a) cylinder b) anchor ring c) Dupin cyclide.

There are phases with the molecular orientation perpendicular to the layer planes and other types with inclined molecules ("tilted phases").

There are several phase types in which the molecules, because of rotational freedom around the long axes, possess rotational symmetry around this axis. In the relatively dense smectic packing the molecules tend to a hexagonal structure, which in the case of the orthogonal structure is called smectic B (Fig. 1.19). Two kinds of smectic B phases can be distinguished. If the positional order within the smectic layers has long range character the "smectic B crystal" phase, also called "smectic L phase" arises. In the so-called "hexatic" smectic B phase only a short-range positional order within the layers exists.

In the case of tilted structures there are several possibilities depending on the direction of the tilt and the positional ordering, smectic I, F, J, G, M (Fig. 1.20).

Whereas the smectic phase F, I and probably M are tilted variants of the smectic B hexatic phase, the tilted smectic types G and J show like the "smectic B crystal" phase long-range positional order within the layers.

In all of the above-mentioned smectic phases there is rotational freedom about the long molecular axes and the structures have hexagonal character in the smectic B and

Fig. 1.16. Smectic A fan-shaped texture.

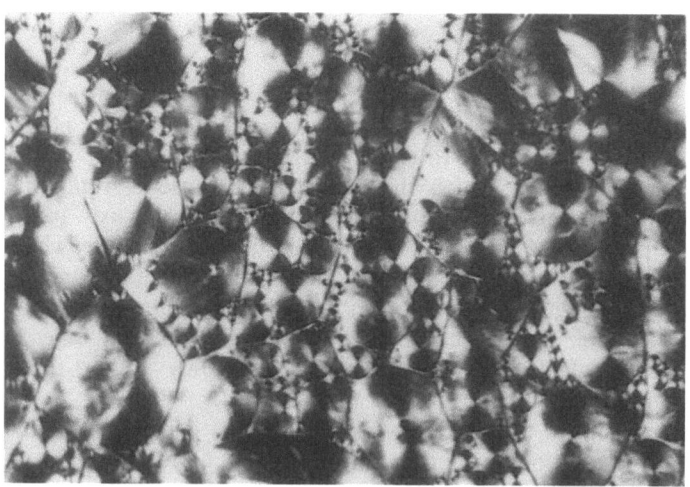

Fig. 1.17. Smectic A polygonal texture.

Fig. 1.18. Smectic C broken fan-shaped texture.

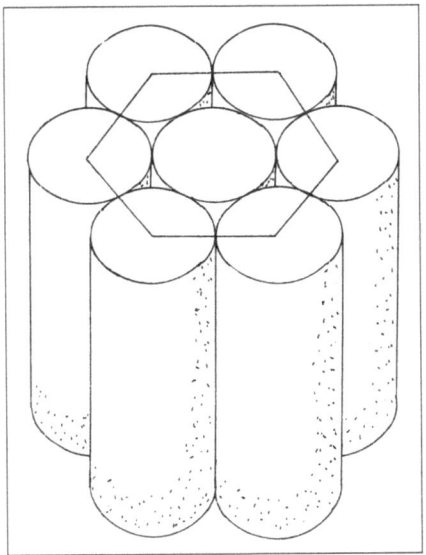

Fig. 1.19. Structure of smectic B phases.

L phases, respectively, "pseudohexagonal" character in the tilted phase types. The term "pseudohexagonal" means that in a plane perpendicular to the molecule director a hexagonal arrangement exists.

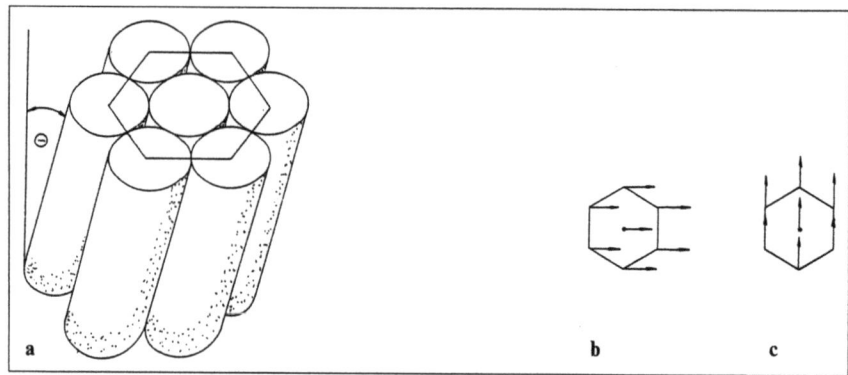

Fig. 1.20. Structure of tilted smectic phases with order within the layers: a) pseudohexagonal structure. Orientation of the molecules in the unit cell; b) tilt to side (smectic F, G); c) tilt to apex (smectic I, J).

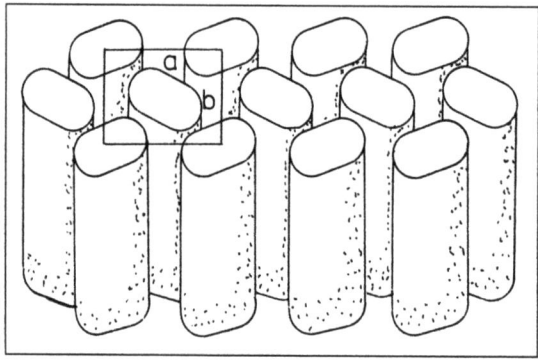

Fig. 1.21. Structure of smectic E phases.

There are additional highly ordered structure variants in which, because of the strong rotational hindrance about the long axes, "herring bone" structures occur. The orthogonal variant of the herring-bone structure is called smectic E (Fig. 1.21). Despite the strong hindrance of the rotational mobility in the structure, dielectric relaxation investigations prove that there is still the possibility of rotations about the long and the short molecular axes.

The tilted herring-bone structure exists in two variants, smectic H and smectic K (Fig. 1.22).

The features of the smectic phases of non-chiral compounds are collected in Table 1.1.

The borderline between the three-dimensional solid crystalline state and the liquid crystalline state is a controversely discussed problem. The "classical" smectic phase types A and C are clearly not three-dimensional with respect to their state of order, however, several highly ordered smectic phase types (e.g., smectic E, H, K) correspond to three-dimensional long-range positional and orientational order. Therefore,

Table 1.1. Structural characteristic of smectic phases of non-chiral compounds.

Phase type	Molecular orientation	Molecular packing	Orientational ordering	Positional ordering	Further features in Fig.
A	orthogonal	random	short range	short range	1.10, 1.11, 1.15, 1.16, 1.17, 1.30, 1.31
C	tilted	random	short range	short range	1.12, 1.13, 1.14, 1.18
M	tilted	random	unknown	short range	
B (hexatic)	orthogonal	hexagonal	long range	short range	1.19, 1.23
I	tilt to apex of hexagon	pseudo hexagonal	long range	short range	1.20, 1.24
F	tilt to side of hexagon	pseudo hexagonal	long range	short range	1.20, 1.25
L (B cryst)	orthogonal	hexagonal	long range	long range	1.19, 1.23
J (G')	tilt to apex of hexagon	pseudo hexagonal	long range	long range	1.20
G	tilt to side of hexagon	pseudo hexagonal	long range	long range	1.20
E	orthogonal	orthorhombic	long range	long range	1.21
K (H')	tilted to side a	monoclinic	long range	long range	1.22
H	tilted to side b	monoclinic	long range	long range	1.22

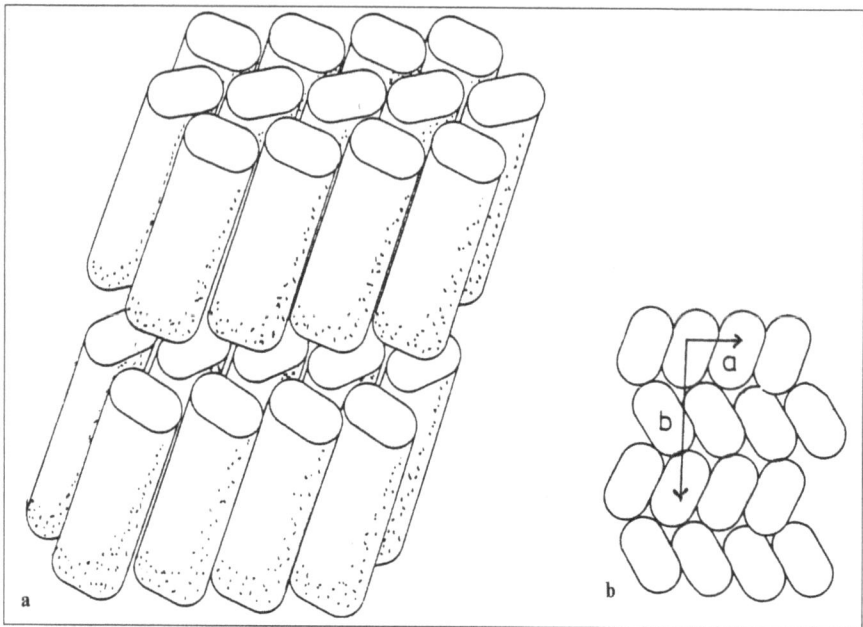

Fig. 1.22. Structure of tilted smectic phases with "herring bone" structure: a) perspective view b) top view of monoclinic unit cells: projection of the molecular packing on to the a–b plane.

the latter are considered by some scientists to be crystalline solids and not liquid crystals. With respect to their molecular mobilities, the textures, elastic and mechanical properties, they are at least closely related to the typical liquid crystals so that, from a more practical standpoint, it seems to be useful to emphasize these strong relations by their designation as liquid crystals.

The ordered smectic phases are able to exhibit many different textures. Some of these textures are typical and may be used for an identification of the phase types, however, there are also unspecific textures which are similar to micrographs of solid crystals. The mosaic texture (Fig. 1.23) is wide spread in ordered smectics. Mosaic textures consist of a conglomerate of small regions which are homogeneously colored, however, the color of neighboring regions usually is different. This points at small liquid "single" crystals in different crystallographical orientation. Mosaic textures are very similar to pictures of solid poly-crystals of metals, minerals and other materials. In several tilted phase types the Schlieren texture (Fig. 1.24) may occur. The Schlieren texture proves the layer structure, including continuous changes of the tilt direction. Because of the high values of the elastic constants such continuous changes are not possible in crystalline solids. Certain ordered smectics are able to exhibit variants of the focal conic textures which are considered to be typical for liquid crystals, e.g., Fig. 1.25 for smectic F.

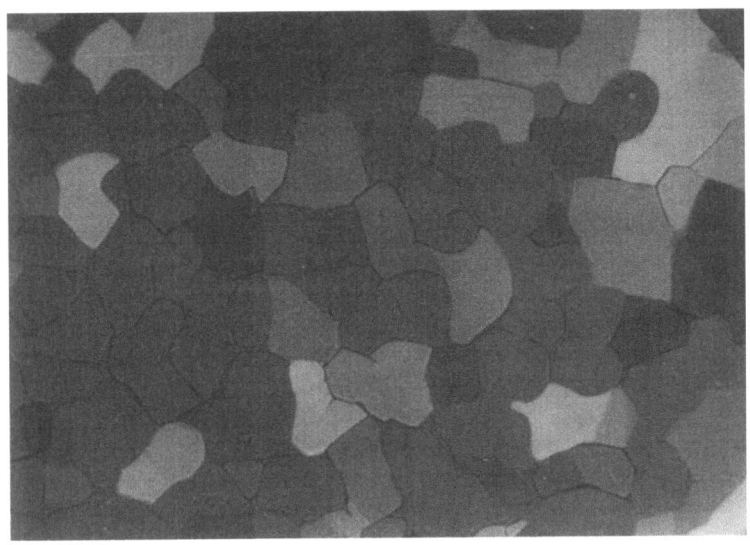

Fig. 1.23. Mosaic texture typical for smectic phases with order within the layers.

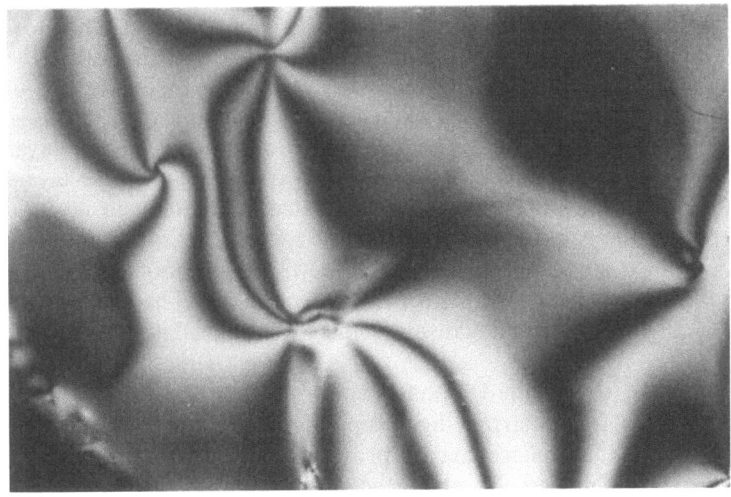

Fig. 1.24. Schlieren texture of smectic I.

Fig. 1.25. Broken fan-shaped texture of smectic F.

Smectic Phases of Chiral Compounds

Many chiral compounds are able to form smectic phases of different types. Some of these phase types (all phases with orthogonal arrangement of the molecules in the layers) have the same structures and properties as the phases formed by non-chiral compounds. In the cases of structures with tilted molecules in the layers, the structures as well as the properties are related, but are distinctly different. The most important smectic phase type derived from chiral compounds is that with the designation C*, which means that this is the analogon to the phase type smectic C of non-chiral compounds. Figure 1.26 gives a sketch of the structure of C* phases. The molecules are tilted in the layers. The projection of the tilt on the layer plane turns from one layer to the next so that a twisted arrangement of the layers arises. The relation of C* to C phases is similar to that of cholesteric to nematic phases. In the C* phases the rotational mobility of the molecules around the long axes is somewhat hindered so that, due to lateral electric dipole moments of the molecules these phases possess ferroelectric properties (Section 2.7). These properties are very promising for use in fast switching displays, this will be elucidated in Chapter 6 in detail.

Because of the twisted structure the C* phases show strong optical activity and selective reflection of circularly polarized light (similar to cholesteric phases; see above). The close relation to the C phases may be proved by the fact that the twist can be enrolled by electric or magnetic fields or by the aid of strong coupling with the surrounding walls. Since the lateral dipoles are cancelled in the twisted arrangement of the C* phases, the ferroelectric properties can be detected only in the enrolled form of the structure.

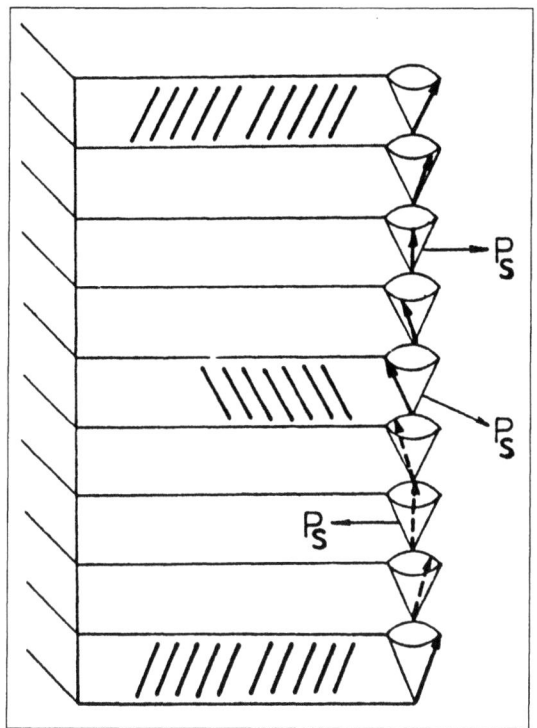

Fig. 1.26. Structure model of the chiral smectic C phase. P_s is the direction of the spontaneous polarization in a smectic layer which is parallel to the layer plane and perpendicular to the tilt plane.

Table 1.2. Ferroelectric smectic phases of chiral compounds.

Random molecular packing

C*	helix, optical activity, selective reflection (Fig. 1.26)
M*	helix, optical activity, selective reflection

Pseudo-hexagonal structure

I*	tilted to side	no layer correlation, short-range in-plane positional order
F*	tilted to apex	helix structure, optical activity, selective reflection
J*	tilted to apex	long-range layer correlation, long-range in-plane positional order
G*	tilted to side	no helix structure

Herring-bone molecular packing

K*	tilted to side	long-range layer correlation, long-range in-plane positional order
H*	tilted to apex	no helix structure

There are several other structures in chiral compounds which are related to the non-chiral analoga. Table 1.2 presents an overview on the most important features of these phases. In principle, also the other ferroelectric smectic phase types are potentially useful for displays, however, because of the problems in homogeneously orienting these phases, up till now the practical use could not be realized.

1.4.1.2   Sequence Rule and Reentrant Behavior

The polymorphism of the crystalline solid state cannot be predicted. Especially there is no law for the sequence of phases of different structures resp. symmetry on the temperature scale.

In the liquid crystalline state, H. Sackmann derived the rule of the phase sequence by systematic observation of the sequences of different phases in polymorphic compounds. This rule in polymorphic compounds predicts a stepwise decrease of order with increasing temperature and the reverse with decreasing temperature. That means, according to this rule, the smectic phases generally are low-temperature phases with respect to the nematic phases. Within the smectic state with increasing temperature first the "herring-bone" structures, then the layer structures with less pronounced positional order and at higher temperatures layer structures without positional order appear. Considering all structures known in calamitic substances, a hypothetical phase sequence can be derived:

solid crystal–H–K–E–G–J–F–L–I–B_{hex}–M–C–A–N–isotropic.

There is no single substance which exhibits all of these phases. At present, smectics with up to five different smectic and one nematic phase are known. Over all, there are many real phase sequences which may be considered as parts of the above given hypothetical full sequence. The rule of the phase sequences has been proved to be valid in many thousands of examples.

In the last decade important exceptions from this rule have been detected. In compounds with association behavior or compounds with several flexible chains (in general, compounds which cause effective "molecule polymorphism" in dependence of the temperature) certain phase sequences may be reversed, a situation called "reentrant" behavior. There are compounds with the sequence N_{re}-S_A-N-is or N- S_A-is; even compounds with double reentrance N_{re}-S_{Are}-N_{re}-S_A-N-is. Generally, the reentrant phenomenon is relatively rare and may be considered as an exceptional behavior.

1.4.1.3   Chemistry of Calamitic Liquid Crystals

At present, more than 20 000 liquid crystalline compounds are known; most of them have rod-like molecular shape and therefore belong to the calamitic liquid crystals. In principle, all compounds which are anisotropic in their molecular shape can form liquid crystalline phases. In order to prove the liquid crystalline properties, however, it is necessary that they have low enough melting temperatures. If the melting temperature of a compound is higher than the clearing temperature, this proof will be a question of the possibility of supercooling the isotropic melt without crystallization.

In many cases this is not possible, and therefore these compounds usually are considered to be not mesogenic.

In order to obtain compounds with rod-like molecular shape and low enough melting temperatures the synthesizing chemists found many possibilities. In most cases the compounds contain ring systems of the aromatic, heterocyclic and alicyclic kinds, which must be linked in such a manner to give molecules with nearly linear shape. But there are a few examples where acyclic compounds show liquid crystalline behavior.

In Table 1.3 some examples for rod-like mesogenic compounds are presented. No. 1 is an acyclic compound. No. 2 is a purely aromatic compound with a rigid exact linear molecule. In many cases the mesogenics are longitudinally substituted by flexible chains, e.g., No. 3. Table 1.4 presents other flexible chains and moieties which may serve as lateral groups. Example No. 4 proves that also cyclohexane rings are well suited for the synthesis of liquid crystals. Many compounds consisting of a middle group which links two terminally substituted ring systems. A typical example is given in No. 5. Other possibilities for middle groups are collected in Table 1.5. One of the most effective bridging groups is the azomethine group, e.g., No. 6, which gives rise to a very high degree of polymorphism. The ring systems can have more than six atoms (No. 7; for other possibilities see Table 1.7) or less than six atoms (No. 8; for further possibilities see Table 1.8). In the last two decades many heterocyclic ring systems have been introduced into liquid crystal chemistry, an example is given in No. 9 and different heterocyclic rings are collected in Tables 1.6, 1.7 and 1.8. There are relatively few compounds with metal atoms in the center, e.g. No. 10. No. 11 is an interesting example for a compound which, due to hydrogen bonds, is dimerized (like all carboxylic acids) and only because of that has the elongated shape needed for mesomorphism. The last compound, No. 12, is one of the cholesteryl esters in which the liquid crystalline state was detected in 1888.

By combination of the moieties compiled in Tables 1.4–1.8, thousands of liquid crystalline compounds can be obtained. The combinations can be made arbitrarily, provided the elongated shape of the molecules is obeyed.

Table 1.9 demonstrates the effect of enlarging the aromatic core of a mesogen. The melting temperatures are strongly enhanced by doing this, but to an even larger extent the clearing temperatures increase. That means that low melting compounds, as they are needed for practical applications, can be obtained especially in systems with small molecular cores. The attachment of flexible terminal groups usually lowers the melting and clearing temperatures of mesogenic compounds. According to a rule already derived by Vorländer in 1908, the mesogenic molecules should have an utmost linear shape. This molecular shape is disturbed by lateral branches, as Table 1.10 demonstrates. By increasing size of the lateral substituent, the nematic clearing temperatures decrease, and even to a higher degree the existence range of the smectic phase.

The synthesis of homologous series provides valuable insights into the connection between molecular structure and properties. Figure 1.27 presents a graphical plot of the transition temperatures versus the length of the alkyl chains in a homologous series. Clearly, the alternation of the temperatures of the transition nematic/isotropic can be seen. This alternation occurs in most of the nematic homologous series. The transition temperatures smectic/nematic show a regular trend within the series which can be considered as a special case of the rule of the regular trend of the transition temperatures in mesogenic series, except the melting temperatures which always are more or less irregular (see also Fig. 1.27). This regular trend of the transition temperatures is an important criterion for the classification of the different liquid crystalline phase types within homologous series.

Table 1.3. Rod-like mesogenic compounds.

1. $CF_3 (CF_2)_9-(CH_2)_9-CH_3$
   (perfluorodecyl)-decane
   cr 38 S_B 61 is

2. ⌬-⌬-⌬-⌬-⌬-⌬ 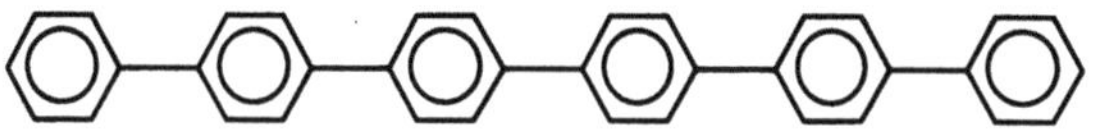
   p-sexiphenyl
   cr 435 S 465 N 565 is

3. C_8H_{17}-⌬-⌬-⌬-⌬-C_8H_{17}
   4,4''-bis-(n-octyl)-p-terphenyl
   cr 176 S 191 is

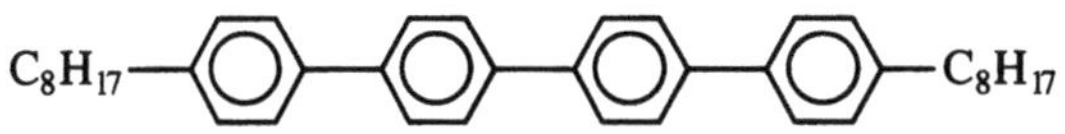

4. $C_{12}H_{25}$-⌬-⌬-⌬-$C_{12}H_{25}$
   4,4''-bis-(n-dodecyl)-p-tercyclohexan
   cr 67 S 185 is

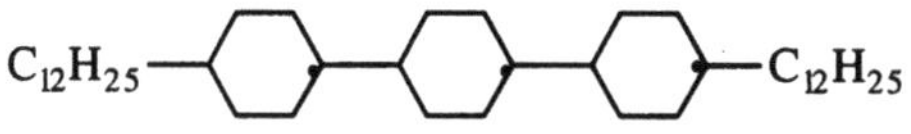

5. C_3H_7-⌬-N=N-⌬-OOC-C_6H_{13}
   4-n-propyl-4'-n-heptanoyloxy-azobenzene
   cr 42 N 71 is

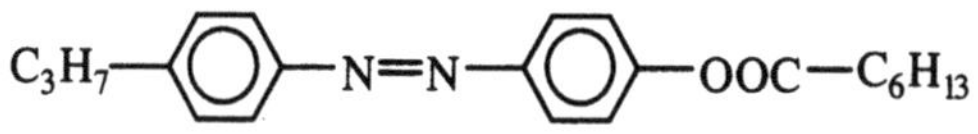

6. $C_6H_{11}O$-⌬-CH=N-⌬-C_7H_{15}
   4-n-heptyl-N-(4-n-pentyloxy-benzylidene)-aniline
   cr 29.5 S_G 33.9 S_B 51.0 S_C 53.1 S_A 62.8 N 78.0 is

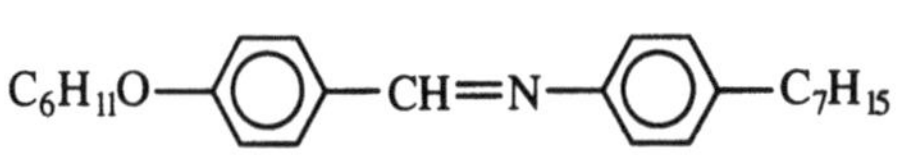

Table 1.3. contd.

---

7. 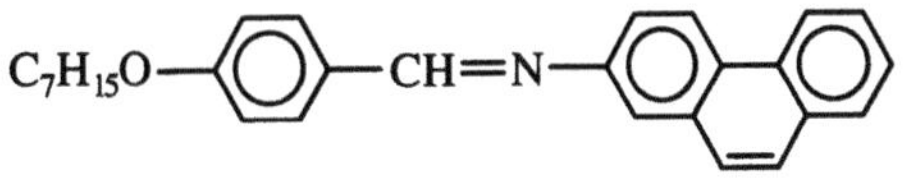 $C_7H_{15}O-\text{C}_6H_4-CH=N-\text{phenanthrene}$

    2-(4-n-heptyloxy-benzylideneamino)-phenanthrene

    cr 116.5 S 129.5 N 187

8. 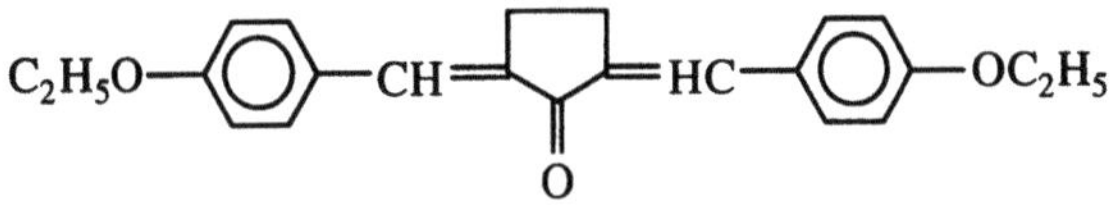 $C_2H_5O-\text{C}_6H_4-CH=\text{(cyclopentanone)}=HC-\text{C}_6H_4-OC_2H_5$

    2,5-bis-(ethoxy-benzylidene)-cyclopentanone

    cr 194 N 202 is

9. 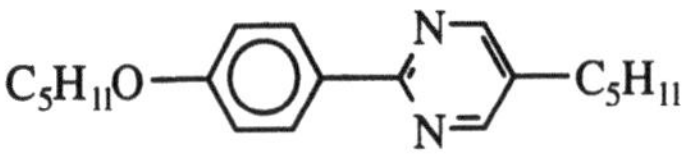 $C_5H_{11}O-\text{C}_6H_4-\text{(pyrimidine)}-C_5H_{11}$

    5-n-pentyl-2-(4-n-pentyloxy-phenyl)-pyrimidine

    cr 37 N 55.5 is

10. 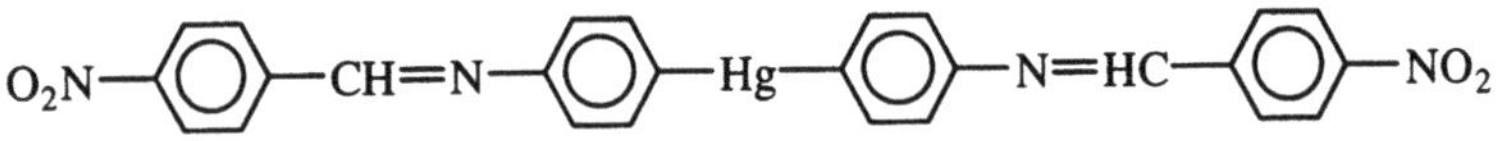 $O_2N-\text{C}_6H_4-CH=N-\text{C}_6H_4-Hg-\text{C}_6H_4-N=HC-\text{C}_6H_4-NO_2$

    bis-(4-nitro-benzylideneamino)-phenyl-mercury

    cr 236 N 241 is

11. 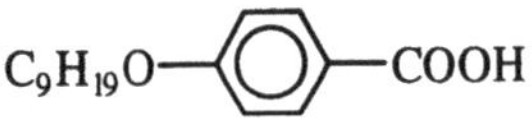 $C_9H_{19}O-\text{C}_6H_4-COOH$

    4-n-nonyloxy-benzoic acid

    cr 94 S_C 117 N 143 is

12. 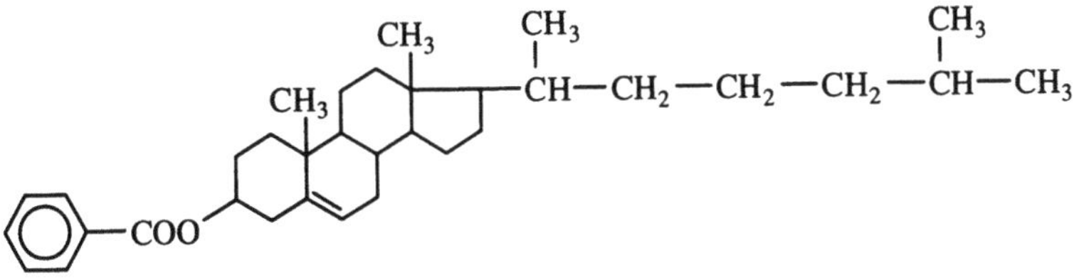 3-β-benzoyloxy-cholesten-(5) (cholesteryl benzoate)

    cr 150.5 N* 182.6 is

Table 1.4. Terminal groups.

alkyl	$-C_nH_{2n+1}$
alkyloxy	$-OC_nH_{2n+1}$
alkylmercapto	$-S-C_nH_{2n+1}$
acyl	$-\underset{\underset{O}{\parallel}}{C}-C_nH_{2n+1}$
acyloxy	$-O-\underset{\underset{O}{\parallel}}{C}-C_nH_{2n+1}$
alkylester	$-\underset{\underset{O}{\parallel}}{C}-O-C_nH_{2n+1}$
alkylcarbonates	$-O-\underset{\underset{O}{\parallel}}{C}-O-C_nH_{2n+1}$
halogeno	$-F, -Cl, -Br, -I$
cyano	$-CN$
nitro	$-NO_2$
alkylamino	$-NH-C_nH_{2n+1}$
cyanoalkyl	$-(CH_2)_n-CN$
cyanoethenyl	$-CH=CH-CN$
dicyanoethenyl	$-CH=C(CN)_2$

Table 1.5. Bridging groups.

azomethine	—CH=N—
ester	—COO—
thioester	—COS—
ethyl	—CH$_2$—CH$_2$—
butyl	—CH$_2$—CH$_2$—CH$_2$—CH$_2$—
stilbene	—C=C—
tolanes	—C≡C—
azo	—N=N—
azoxy	—N=N— ↓ O
oximbenzoate	—C=N—OOC— |
dicarboxylate	—OOC—(CH$_2$)$_n$—COO—
azine	—CH=N—N=HC—
mercury	—Hg—

Table 1.6. Ring systems with 6 atoms.

phenyl	
cyclohexane	
cyclohexanone	
piperidine	
pyridine	
piperazine	
dioxane	
dithiane	
oxathiane	
pyridazine	
pyrimidine	
pyrazine	
triazine	
tetrazine	

Table 1.7. Ring systems with more than 6 atoms.

biphenyl	
naphthalene	
decaline	
fluorene	
perhydrophenanthrene	
dioxanaphthalene	
bicyclooctane	
bicyclotrioxane	
cubane	

Table 1.8. Ring systems with 5 atoms.

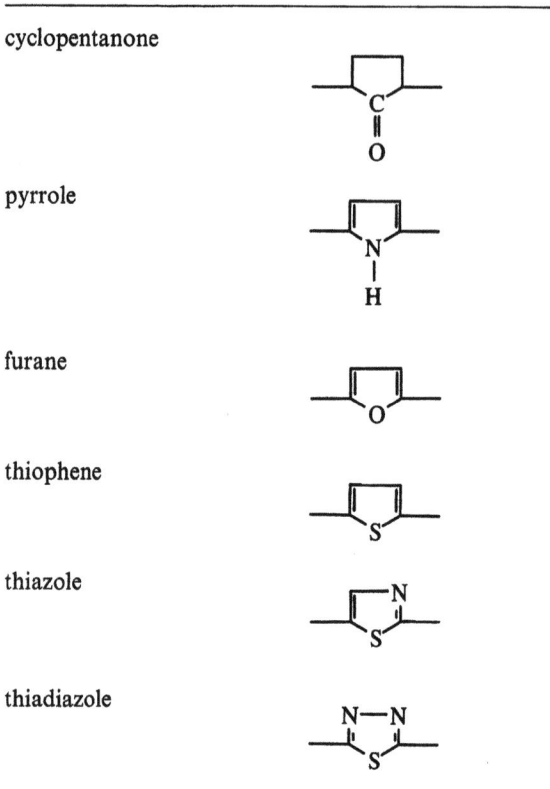

cyclopentanone	
pyrrole	
furane	
thiophene	
thiazole	
thiadiazole	

Table 1.9. Melting and clearing temperatures in dependence of the length of the aromatic core.

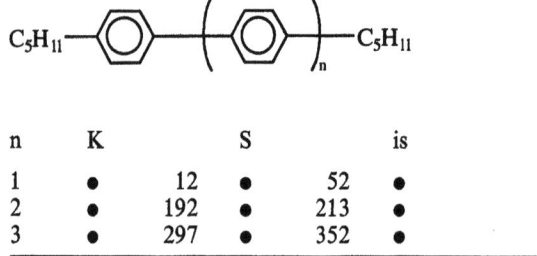

n	K		S		is
1	●	12	●	52	●
2	●	192	●	213	●
3	●	297	●	352	●

Table 1.10. Transition temperatures in dependence of the lateral substituents.

H₅C₂O—⟨◯⟩—CH=N—⟨◯⟩—CH=C—COOC₂H₅
                                               |
                                               R

R	cr		S		N		is
H	●	81	●	157	●	160	●
CH$_3$	●	95	●	(77)	●	123	●
C$_2$H$_5$	●	73	–	–	●	(62)	●
C$_6$H$_5$	●	104	–	–	–	–	●

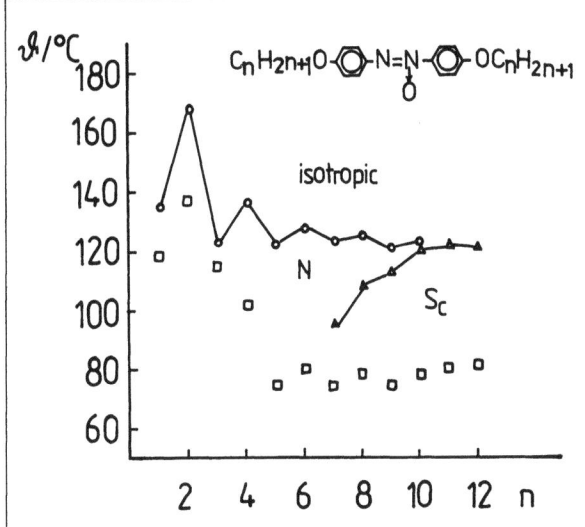

Fig. 1.27. Transition temperatures in the homologous series of 4,4'-di-n-alkyloxy azoxybenzenes (n: number of carbon atoms in the terminal alkyl chains)

1.4.1.4 Molecular Statistical Theories of the Calamitic Phases

The liquid crystalline state may be considered as a continuum and, in this sense, treated by theories using elastic properties. This type of theoretical description is important for calculating electrooptical properties of liquid crystals and, therefore, the basis for a theoretical treatment of liquid crystal displays (see Chapter 6). The connection of the molecular properties with the macroscopic properties of liquid crystals can be achieved by using molecular statistical methods. The molecular statistical treatment usually starts from the nematic state since this is the simplest liquid crystalline state. Theories of the smectic states are obtained by taking additional order parameters to the theory of nematic state.

Molecular Statistical Theory of the Nematic State

As already mentioned above the main difference of the nematic state to isotropic liquids is the average parallel orientation of the molecular long axes. The intermolecular mobility is about the same in both phases.

It has been a long discussed question which intermolecular forces are responsible for the stabilization of nematic phases. Theoretical and experimental results of the last 15 years have elucidated this question in principle.

The theoretical description of the nematic state is based on a comparison with the corresponding isotropic liquid. The free energy F, which at constant volume and temperature is a measure for the stability of a phase, may be expressed in the frame of statistical theories in terms of the partition function Q_C:

$$F = -kT \ln Q_C$$

k = Boltzmann factor T = temperature (Kelvin)

$$Q_C = \frac{1}{N!} \int d\Omega_i \int dR_i \exp(-U_C/kT).$$

Q_C is restricted here to that part of the inner energy U_C (configurational part) which depends on the configuration (orientation Ω_i and distance R_i) of the molecules. The translational, vibrational, and rotational parts of the inner energy are assumed to be equal in the nematic and the isotropic state. U_C may be expressed as a sum of pair potentials u_{ij}:

$$U_C = \sum u_{ij}.$$

Pair potentials generally can be split into an attractive (u_{ij}^{attr}) and a repulsive (u_{ij}^{rep}) part:

$$u_{ij} = u_{ij}^{attr} + u_{ij}^{rep}.$$

The attractive term may be divided into two parts which consider an isotropic attraction which also exists in isotropic liquids, and an anisotropic attraction which is specific for the nematic state. Because of the mathematical difficulties in an exact theoretical description of nematic liquids simplified models must be used. One of the most effective models is the mean field description, that means an averaged molecular interaction with all neighbors as it is "felt" by a molecule. Within the mean field approximation the attractive potential may be expressed as

$$V(\beta, \delta) = -L_o V_o \delta - L_2 V_o \delta\, S\, (3\cos^2\beta - 1)/2$$

where

δ = number density of molecules
V_o = molar volume

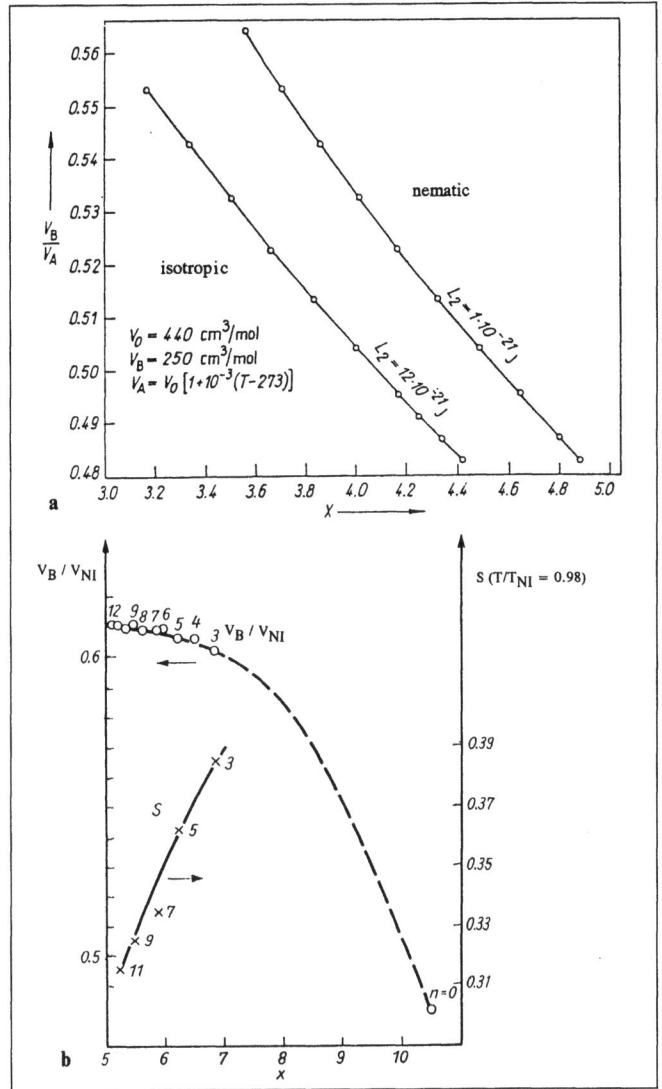

Fig. 1.28a. Packing fraction V_B/V_A versus length-to-breadth ratio X for two anisotropic attraction parameters L_2 according to the van der Waals theory of Cotter [2]

Fig. 1.28b. Packing fraction V_B/V_{NI} at the clearing temperature (T_{NI}) and orientational order parameter $S(T/T_{NI} = 0.98)$ in dependence of the length-to-breadth ratio X for the homologous 2-n-alkyl-p-phenylene-bis-(4-n-octyloxybenzoates). The numbers at the curves indicate the number of carbon atoms in the lateral chains.

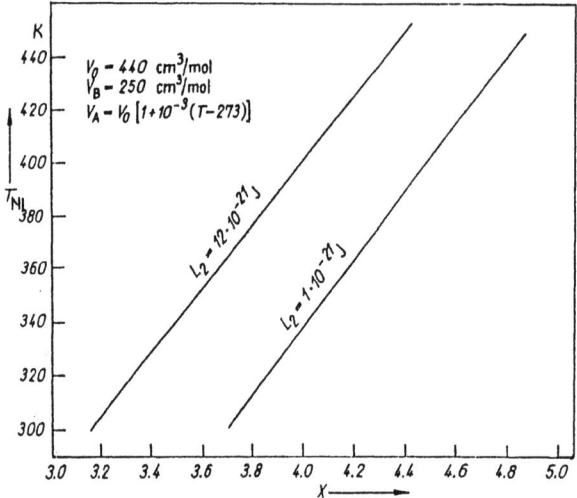

Fig. 1.29. Clearing temperature T_{NI} versus length-to-breadth ratio X for two anisotropic attraction parameters L_2 [2].

β = angle of the molecular long axis with respect to the director
S = orientational order parameter
L_o = isotropic attraction parameter
L_2 = anisotropic attraction parameter.

The molecules are approximated by rigid hard cylinders which are coated at both ends by half-spheres. The cylinders are characterized by the length-to-breadth ratio X:

$$X = \frac{L}{2W} + 1,$$

where L is the length and W the breadth of the molecule.

The cylinders can come near together, but cannot overlap each other which is expressed in terms of a repulsive "hard rod potential":

$$u^{\text{rep}} = \begin{cases} \infty & \text{if the cores i and j would overlap} \\ 0 & \text{otherwise} \end{cases}.$$

When V is the actual volume and V_B is the fictive, most dense packing (calculated by an increment system of Bondi [1]) the packing fraction V_B/V can be calculated as a measure for the space which is filled with material. At low packing fractions the system is isotropic. In the van der Waals theory [2] a critical packing fraction can be calculated above which the system becomes anisotropic nematic (Fig. 1.28a). In terms of this theory the structure of the nematic liquid is mainly determined by the repulsion; the attraction is mainly necessary in order to obtain the required critical density. Especially the anisotropic part of the attraction connected with L_2 (which are mainly considered in the theory of Maier and Saupe [3]) plays a minor role.

Attractive forces are determined mainly by dispersion interaction, and polar forces in the most cases are much less dominant. Figure 1.28b presents experimental data qualitatively confirming the theory.

The theory also allows the calculation of the temperatures of the transition nematic/isotropic (Fig. 1.29). In this presentation the dominance of the form anisotropy in determining the clearing temperatures is clearly recognizable.

The theoretical considerations allow important conclusions concerning the connection of the molecular structure and the properties of nematic liquid crystals, which is specially important for the derivation of synthesis concepts of liquid crystals (see also Section 1.4.1.3).

Molecular Statistical Theory of the Smectic State

McMillan [4] extended the Maier-Saupe theory of the nematic state (which is based on the orientation-dependent part of dispersion forces) to the smectic A phase. He included an additional principle of order which describes the one-dimensional translational periodicity of the layer structure.

The molecules are considered to have a central aromatic part and two flexible terminal alkyl chains. Between the cores there is a strong dispersion interaction because of the large polarizability of the aromatic part. If the alkyl chains are sufficiently long to separate the aromatic cores a layer arrangement results.

From the anisotropic pair interaction energy the one-particle potential a test molecule would feel is given by

$$V_1 = V_o\{\sigma\alpha\tau \cos(2\pi z/d) + (S + \sigma\alpha \cos(2\pi z/d))(3\cos^2\beta - 1)/2\},$$

where
  z = co-ordinate in direction of the layer normal
  d = thickness of the smectic layer
  β = angle between the long molecular axis and the director
  V_o = strength of the potential
  $S = \langle(3\cos^2\beta - 1)/2\rangle$ orientational order parameter
  $\tau = \langle\cos(2\pi z/d)\rangle$ smectic order parameter; a measure of the amplitude of the density waves describing the smectic layer structure
  $\delta = \langle\cos(2\pi z/d)(3\cos^2\beta - 1)/2\rangle$ order parameter, which couples the orientational with the translational order. The parameter α is defined as

$$\alpha = 2\exp - (r_o z/d)^2,$$

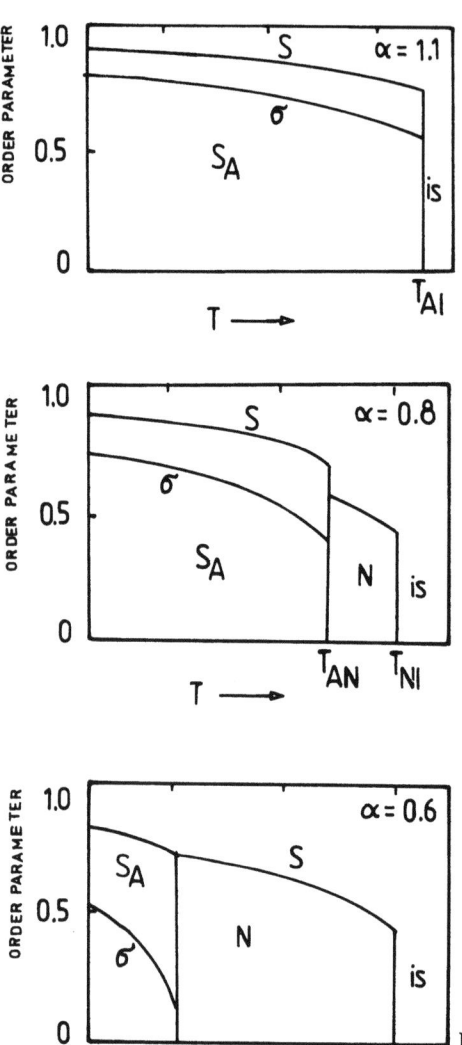

Fig. 1.30. The order parameters S and σ, respectively, versus temperature for three values of the parameter α (1.1; 0.8; 0.6) (accord. [4])

where r_o is the length of the rigid core. This parameter varies from 0 to 2. According to its definition α increases for a given r_o with increasing length of the flexible terminal chains of the molecules.

According to the theory three stable states may occur in dependence of the parameters α and δ

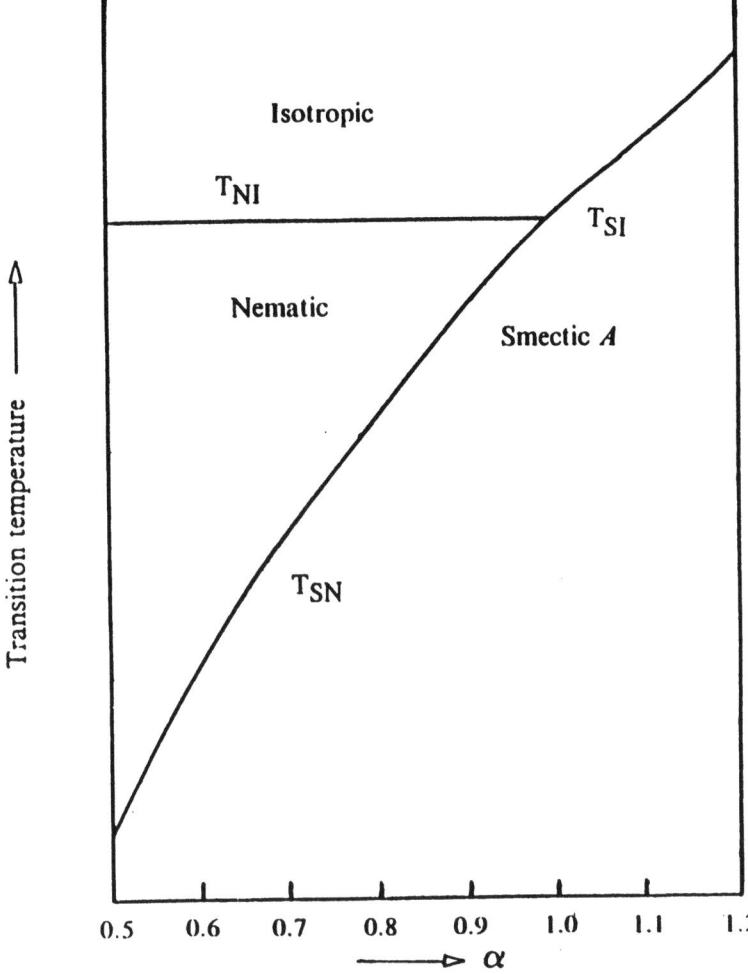

Fig. 1.31. Transition temperatures (T_{NI}, T_{SI}, T_{SN}) in dependence of the parameter α [4].

isotropic liquid $S = 0$ $\tau = 0$ $\delta = 0$
nematic phase $S \neq 0$ $\tau = 0$ $\delta = 0$
smectic A $S \neq 0$ $\tau \neq 0$ $\delta \neq 0$.

It follows from the theory that for $\alpha < 0.7$ a second-order phase transition $S_A \rightarrow N$ results which becomes first-order for $\alpha > 0.7$. For $\alpha \geq 0.98$ a direct transition of the S_A phase into the isotropic liquid occurs (see Fig. 1.30). Fig. 1.31 presents the phase transition behavior as a function of the parameter α, as predicted by the theory.

The McMillan theory is able to describe some properties of smectic A phases in a qualitative manner. Quantitative agreement cannot be expected since the used molecular model is very rough, and the important repulsive forces are neglected. The model has been revised and further developed, however, quantitative agreement cannot be obtained.

By introduction of additional order parameters theories for other types of smectic phases can be elaborated. For example for smectic C phases the tilt angle is used as an additional order parameter. All these theories are based on relatively crude models and therefore are able to describe the reality only in a rough approximation. It is not possible to predict the polymorphism and phase behavior of a given chemical compound on the basis of such theories.

1.4.2. Discotic Liquid Crystals

Mesophases are not only possible in molecules with rod-like shape, but also in form-anisotropic molecules of disc-like shape. In 1977, Chandrasekhar discovered liquid crystalline phases in hexa-substituted benzene derivatives (Table 1.11, No. 1). The molecules of this and other substance classes are flat and disc-like in shape and able to be packed in different structures.

1.4.2.1. Phase Structures of Discotic Liquid Crystals

Similar to rod-shaped molecules, also disc-shaped molecules can form a nematic phase characterized by a long-range orientational order, but no long-range positional order. The director now corresponds to the preferred direction of the disc normals (see Fig. 1.32) This discotic nematic phase (N_D) shows textures typical for nematic phases, e.g., Schlieren textures. In the case of chiral compounds the discotic nematic phase exhibits a twisted structure.

In most cases, the disc-shaped molecules are packed one upon another to form columns. Within the columns the molecules can have a certain order (short designation o) or disorder (short designation d), see Fig. 1.33.

Generally the alkyl chains are assumed to be liquid-like. The columns itself are arranged in a two-dimensional network leading to *columnar phases* with hexagonal, rectangular or oblique symmetry whereby also tilted variants are possible (Fig. 1.33).

Figure 1.34 presents a texture photograph of a columnar phase.

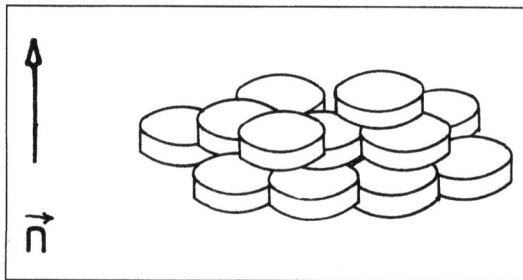

Fig. 1.32. Structure model of the discotic nematic phase (N_D) ($\vec{n}$: director).

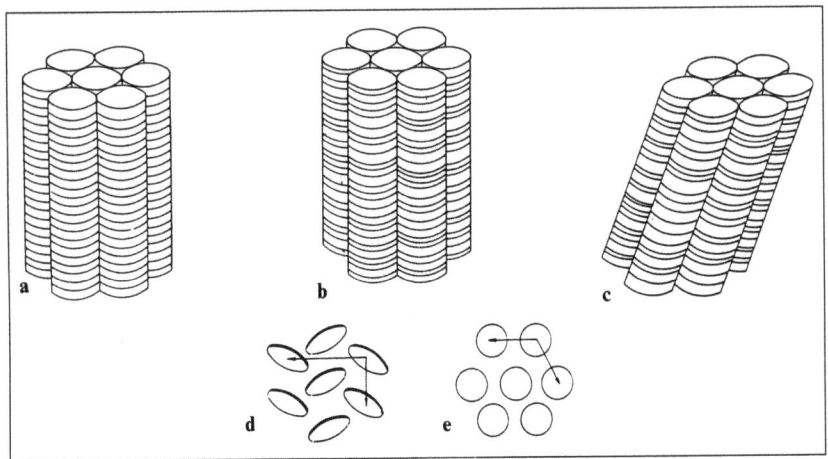

Fig. 1.33. Columnar phases of disc-shaped molecules a) ordered; b) disordered; c) tilted d) rectangular (top view; ellipses denote discs that are tilted with respect to the column axis e) hexagonal (top view))

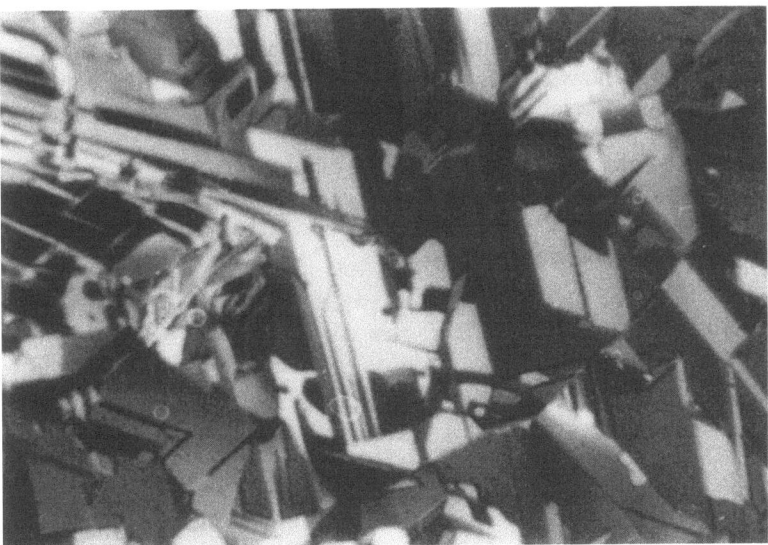

Fig. 1.34. Texture of a hexagonal columnar mesophase

Despite the fact that most of the textures of columnar phases have not been analyzed in detail, there are typical pictures which may be used for the recognition of these phases and, in favorable cases, for the classification of them.

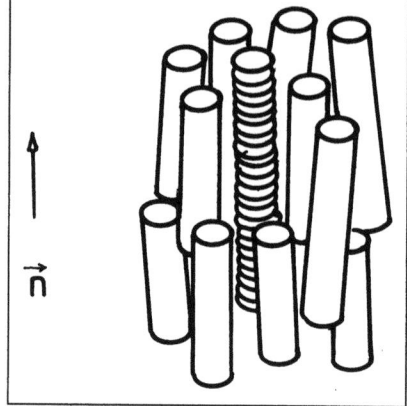

Fig. 1.35. Structure model of the columnar nematic phase ($\vec{n}$: director).

Very recently, in two-component systems, columnar structures have been found in which the columns are not arranged on a regular lattice. The only principle of order is their approximative parallelity. Therefore, they are comparable to nematic structures of rod-like molecules and, in fact, they are columnar nematic phases (Fig. 1.35).

Some of the disc-like compounds exhibit only one liquid crystalline phase, other compounds exhibit several and up to four different liquid crystalline phases. In certain cases the discotic nematic phases exist at higher temperatures than the columnar phases, in other compounds the reverse is true. Generally, in discotic compounds a rule of the phase sequence as it has been found for calamitic phase does not exist.

1.4.2.2 Chemistry of Discotic Liquid Crystals

Discotic compounds consist of flat, rigid cores which are surrounded by flexible chains (alkyl, alkyloxy, alkanoyloxy). There are compounds with 3,4,6,8,9 or 12 flexible chains. The effect of the latter is to lower the melting temperature and to isolate the columns one to another so that a structure with some mobility can be formed.

Typical examples are compiled in Table 1.11. Number 1 is the compound in which the discotic phase has first been observed. Number 2 proves that also compounds without long flexible chains can exceptionally form columnar structures. Compound No. 7 should be considered with special interest since the core of this substance is not flat, but rather cone-shaped. The substance class derived from cores with that property is therefore called "bowlic" or "pyramidic". These compounds are able to form columnar phases. Due to the special shape of the molecules, in order to obtain a dense packing, the up and down symmetry existing in ordinary discotic substances is no longer preserved. In columns of such compounds the eventual electrical partial dipole moments in the direction of the column long axis should add to a kind of ferroelectric column. Also, anomalously high electrical conductivity should be possible in structures of this type, however, detailed investigations are still lacking.

In the literature macrocyclic systems have been described which have holes in the center. These compounds are able to build columnar phases which are called "tubular mesophases".

Table 1.11. Typical molecular structures of discotic liquid crystals.

1

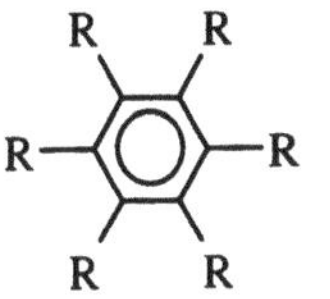

—R = C_7H_{15}—COO—

cr 79.8 D 83.4 is

2

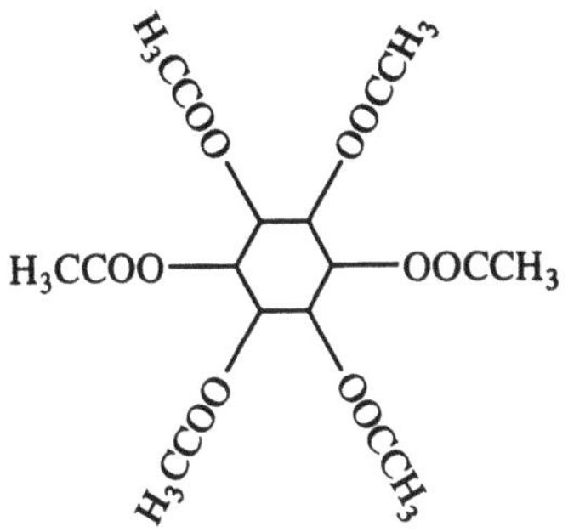

cr 288 D 293 is

3

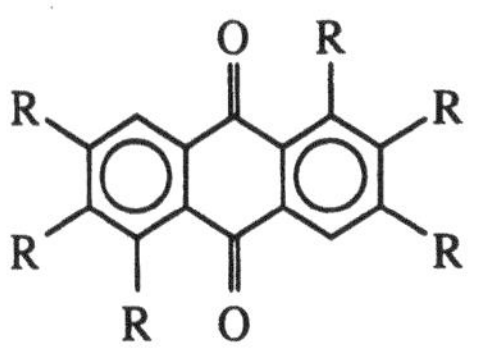

R = C_7H_{15}—COO—

cr 107 (D_2 95) D_1 127 is

4

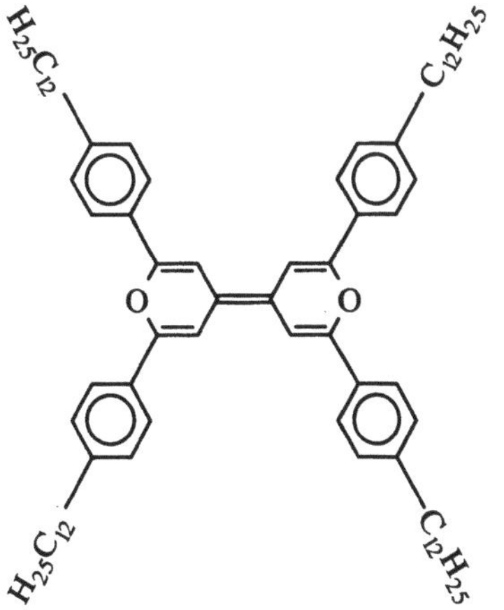

96 D 147 is

Table 1.11. contd.

5

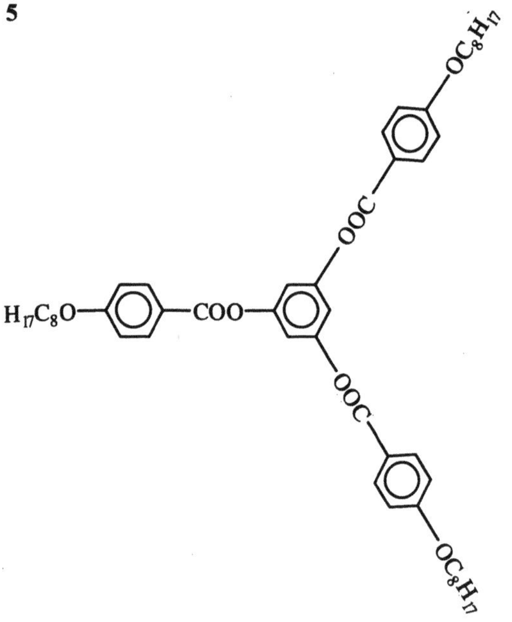

H₁₇C₈O—⌬—COO—⌬(—OOC—⌬—OC₈H₁₇)(—OOC—⌬—OC₈H₁₇)

72.7 (N_D 23.5) is

6

[porphyrazine-like macrocycle with four benzo units bearing R groups]

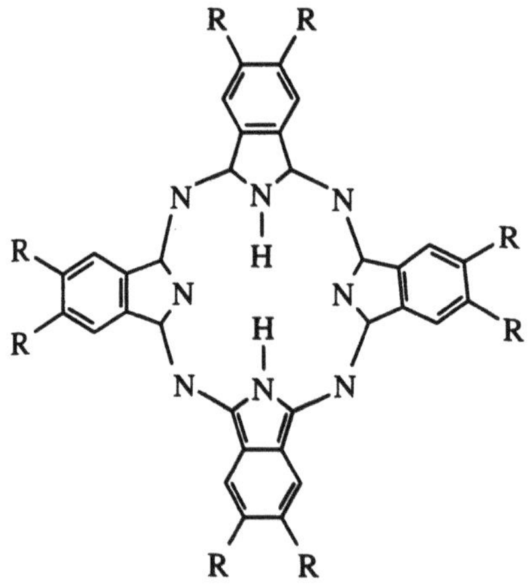

R = -CH₂OC₁₂H₂₅

78 D 264 is

7

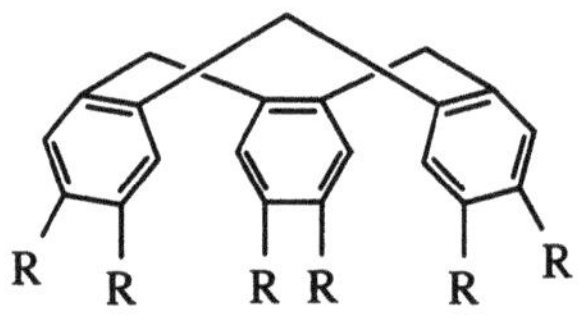

[cyclotriveratrylene-type structure with three aryl rings each bearing two R groups]

R = C₉H₁₉COO—

23.9 columnar 152.6 is

1.4.2.3 Molecular Statistical Theory of Discotic Phases

Madhusudana et al. [5] applied a molecular-statistical theory to a system of hard right circular cylinders where the form anisotropy of the molecules is expressed by the radius (r) to height (l) ratio $R = r/l$. This ratio is > 0.5 for disc-shaped molecules and < 0.5 for rod-like molecules. Madhusudana et al. were able to show that not only rod-like, but also disc-shaped molecules can form nematic phases. Their theory is based only on the action of repulsive forces; attractive forces are neglected. As it turns out in qualitive agreement with the experimental results—with increasing form anisotropy of the molecules the mesogenity increases, which is indicated by a decreasing value of the critical packing fraction at the transition nematic-isotropic (that means the dense packing which is necessary to stabilize the anisotropic phase with respect to the coexisting isotropic state).

Although the theory is based on a very rough model, qualitatively the existence of nematic phases for calamatic as well as discotic materials can be understood.

1.4.3 Sanidic Liquid Crystals

In the preceding sections rod-like and disc-like molecules have been discussed in view of their mesogenic properties. Both these molecular types are special cases of form-anisotropic molecules. These special cases are characterized by the rotational symmetry of the used molecular models, which means the anisotropy of the models can be described by two main axes. The more general case would be the consideration of molecular models with three different main axes.

Molecules of this kind could be characterized by boards. Depending on the relative size of the main axes, these molecules can be derived from rod-like or disc-like molecules. Recently, in fact, it has been shown that compounds of this kind can exhibit mesomorphic properties.

Derived from the Greek word for board, these phases are called "sanidic". Sanidic phases were first found in polymeric liquid crystals.

According to the rule of Vorländer, the mesogenic compounds should have the utmost linear shape. The experimental results proved that lateral branches diminish the mesogenity; large branches in this sense should completely destroy the mesogenic properties. In recent years, however, a number of compounds has been found in which very long chains at the end of the molecules (No. 1 in Table 1.12) or attached in the center (No. 2 in Table 1.12) diminish the transition temperatures mesophase/isotropic phase, but absolutely do not prevent liquid crystalline behavior. The explanation for this unexpected behavior is that the flexible alkyl chains adopt such a conformation that they are nearly parallel to the basic molecule. Despite the fact that they are not exactly rod-like, they exhibit the same types of mesomorphic phases as do the ordinary rod-like compounds.

If the breadth of such molecules is even more enlarged, as the following example shows,

$R = C_6H_{13}$

cr 51.2 N_b 59.6 is

2,3,4-tris-hexyloxy cinnamic acid

Table 1.12. Non-conventional liquid crystals

$$\begin{array}{c} H_{11}C_5OOC \\ \phantom{H_{11}C_5OOC}\diagdown \\ C=CH-\bigcirc-OOC-\bigcirc-\bigcirc-COO-\bigcirc-CH=C \\ \phantom{H_{11}C_5OOC}\diagup \\ H_{11}C_5OOC \end{array} \begin{array}{c} \diagup COOC_5H_{11} \\ \\ \\ \diagdown COOC_5H_{11} \end{array}$$

1   4,4' bis-(di-n-pentyl 4-carbonyloxybenzylidene malonate)-biphenyl
        cr 95 (S_C 61) N 151 is

$$C_8H_{17}O-\bigcirc-COO-\bigcirc-OOC-\bigcirc-OC_8H_{17}$$
$$|$$
$$C_{12}H_{25}$$

2   1,4-bis-(4-n-octyloxy-benzoyloxy)-2-n-dodecyl-benzene
        cr 56.2 N 69.2 is

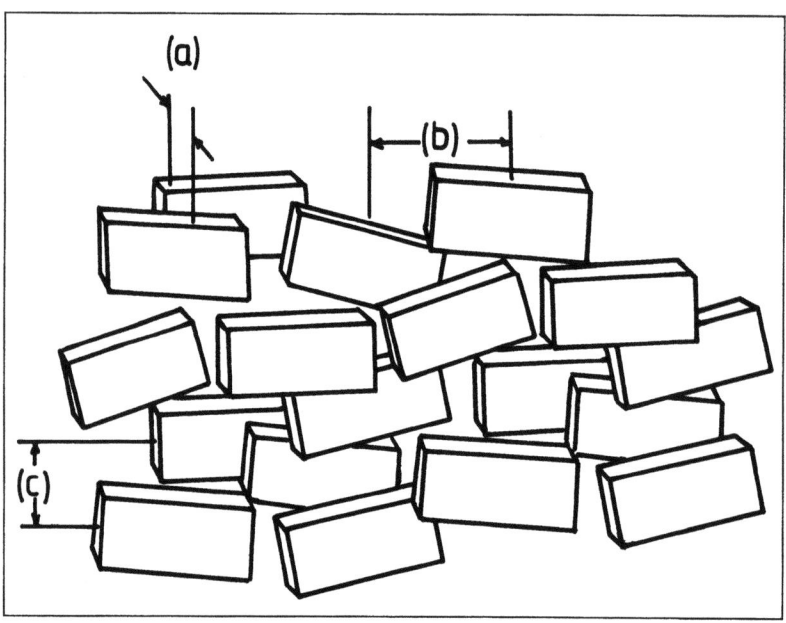

Fig. 1.36. Structure model of the sanidic nematic phase; a, b, c are the lattice parameters according to x-ray data

liquid crystalline properties may still exist, but the phase type is changed. The board-like molecule of this compound can form a nematic phase, however, contrary to the behavior in "normal" nematic phases, the rotation around the molecular long axis is strongly hindered.

Therefore, the structure of these phases (Fig. 1.36) is characterized by three translational periods (in accordance with length, breadth, and thickness of scattering maxima which cause three the molecular units by investigation with x-rays. From the structure model shown in Fig. 1.36 biaxiality could be expected. But up to now an unambiguous experimental proof is still lacking.

1.4.4 Cubic Mesophases

There are several rod-like compounds which exhibit cubic mesophases. In the solid state, crystals with cubic symmetry can be formed by lattice units (atoms, molecules) of spherical shape. Therefore, the occurrence of cubic mesophases in rod-like compounds is somewhat unexpected and needs special explanation. A cubic mesophase has been found in the following compound:

$C_{16}H_{23}O$—⟨O⟩—⟨O⟩—COOH
          |
         NO_2

cr 126.8 S_C 171 cubic 198.5 S_A 199.8 is
4'-n-hexadecyloxy-3'-nitro-biphenyl-4-carboxylic acid

As is normal for carboxylic acids, because of hydrogen bonds the molecules exist as dimers. The smectic A and C phases do not show peculiarities compared with those of other rod-like compounds. In the indicated temperature region between these two smectic phases with layer structure an optically isotropic phase exists. Its structure has been derived on the basis of x-ray investigations which indicate a cubic lattice. Since the lattice parameter is essentially greater than the molecular length, the lattice units seem to be of micellar nature. The most probable assumption is that there are bundles of molecules which may rotate as a whole, thus allowing spherical symmetry.

As x-ray investigation proves, there is three-dimensional long-range order with respect to the position of the lattice units. On the molecular level, however, the structure is liquid-like with "molten" alkyl chains, as a blurred halo in the large-angle region indicates.

The viscosity of the cubic phase is very high, presumably because of the three-dimensional structure. The transitions connected with the cubic phase can be supercooled remarkably. Such behavior is rare in phase transitions of mesophases.

There are a few other compounds of different substances classes which exhibit cubic thermotropic mesophases. The reason for the occurrence of such cubic phases is not clear.

Cubic mesophases are also known in the case of lyotropic mesophases. Here, the structures are discussed in terms of complicated interwoven networks (see Chapter 4).

References

1. Bondi A (1964) Van der Waals Volumes and Radii. J Phys Chem 68:441
2. Cotter MA (1977) Hard spherocylinders in an anisotropic mean field: A simple model for a nematic liquid crystal. J Chem Phys 67:1098
3. Maier W, Saupe A (1958) Einfache molekularstatistische Theorie der nematischen Phase. Z Naturforsch A 19:564
4. McMillan WL (1971) Simple molecular model for smectics A, Phys Rev A 4:1238
5. Savithrama KL, Madhusudana NV (1981) Scaled Particle Theory of a System of Hard Right Circular Cylinders. Mol Cryst Liq Cryst 74:243

General

1. Kelker H, Hatz R (1980) Handbook of Liquid Crystals. Verlag Chemie, Weinheim
2. Bahadur B (ed) (1990) Liquid Crystals. Applications and Uses, vols 1, 2, 3. World Scientific, Singapore New Jersey London Hong Kong

Physics

3. Gennes PG de (1974) The Physics of Liquid Crystals. Clarendon, Oxford
4. Vertogen G, Jeu WH de (1980) Thermotropic Liquid Crystals-Fundamentals. Springer, Berlin-Heidelberg-New York

Chemistry

5. Demus D, Demus H, Zaschke H (1974) Flüssige Kristalle in Tabellen. Deutscher Verlag f. Grundstoffindustrie Leipzig
6. Demus D, Zaschke H (1984) Flüssige Kristalle in Tabellen II. Deutscher Verlag f. Grundstoffindustrie Leipzig
7. Demus D (1988) 100 Years Liquid Crystals Chemistry, Mol Cryst Liq Cryst 165:45
8. Demus D (1989) One hundred years of liquid-crystals chemistry: thermotropic liquid crystals with conventional and unconventional molecular structure, Liq Cryst 5:75
9. Gray GW (1987) Thermotropic Liquid Crystals, J. Wiley & Sons, Chichester, New York, Brisbane, Toronto, Singapore
10. Thiem J, Vill V (1992–1994) Landolt-Börnstein, New Series, Group IV, Macroscopic and technical properties of matter, vol. 7, Liquid Crystals. Springer, Berlin-Heidelberg

Polymorphism, Structure, Texture

11. Demus D, Richter L (1978) Textures of Liquid Crystals, 2nd edn. Deutscher Verlag f. Grundstoffindustrie Leipzig
12. Gray GW, Goodby JW (1984) Smectic Liquid Crystals. Leonard Hill, Glasgow-London. Heyden & Son, Philadelphia
13. Demus D, Diele S, Grande S, Sackmann H (1983) Polymorphism in Liquid Crystals. Advances in Liquid Crystals, vol 6, p 1–107, Academic Press, New York–San Francisco–London
14. Chandrasekhar S, Ranganath GS (1990) Discotic liquid crystals. Rep Progr Phys 53:57–84
15. Sackmann, H (1989) Smectic Liquid Crystals. A historical review. Liq Cryst 5:43
16. Kleman M (1977) Points, Lignes, Parois dans les Fluides Anisotropes et les Solides Cristallins, vols 1, 2. Les Editions de Physiques, Orsay

2 Thermodynamic Behavior and Physical Properties of Thermotropic Liquid Crystals

G. Pelzl

Introduction

The parallel alignment of anisotropic molecules gives rise to the anisotropy of various physical properties, which is the striking feature of the liquid-crystalline state. It will be shown that there is a close relation between the anisotropy of physical properties and the molecular structure, as well as the phase structure. The anisotropy of physical properties is an essential condition for a number of practical applications besides the temperature range and the chemical stability. Therefore, the knowledge of structure-property relations is of growing interest in molecular engineering of liquid crystalline compounds with specific properties.

2.1 Calorimetric Investigations

An exceptional variety of phase transitions is characteristic of the liquid crystalline state. Commonly, phase transitions in which liquid crystalline phases participate are observed microscopically. But, in some cases, the optical textures of the liquid crystalline phases show minimum differences so that it is difficult or impossible to distinguish different phase types by microscopic observations. A more objective and detailed study of phase transitions can be performed by calorimetric investigations. In most cases, differential scanning calorimetry (DSC) is used, but in some cases also adiabatic calorimetry or differential thermal analysis (DTA) are used. With calorimetry not only the temperatures of the phase transitions, but also transition enthalpies, entropies or molar heat capacities can be measured. The phase transition enthalpies are a direct measure of the energetical change and are due to the structural changes occurring at the phase transitions.

The transition enthalpies for different phase transitions are spread over an interval of more than four orders of magnitude. Generally, there is no correlation between the special transition type and the magnitude of the transition enthalpy, because for the same transition type the transition enthalpy strongly depends on the molecular structure. Therefore, information about the relation between transition enthalpy and transition type is only meaningful if chemically similar compounds, i.e., members of a homologous series, are compared. As a representative example, the phase-transition enthalpies of the homologous terephthalylidene-bis-alkylanilines are listed in Table 2.1. It is seen that the transition from the solid to the isotropic liquid state is performed over a number of liquid crystalline phases of decreasing order, so that we can speak of a "melting of instalment". Some regularities can be generalized from Table 2.1. The melting enthalpies for all members of the series are considerably higher

Table 2.1. Transition enthalpies (J mol^{-1}) of the homologous terephthalylide bis-4-n-alkylanilines.

C_nH_{2n+1}—⟨◯⟩—N=HC—⟨◯⟩—CH—N—⟨◯⟩—C_nH_{2n+1}

n	cr	S_5	S_H	S_G	S_F	S_I	S_C	S_A	N	is
2	· 30100	–	–	·	–	·	–	–	· 620	·
3	· 14300	· 1300	· 600	· 7000	–	·	*	· 385	· 1100	·
4	· 19800	· 410	· 1300	· 4370	–	·	*	· 590	· 1350	·
5	· 15100	–	· 1140	· 95	· 3660	·	*	· 1200	· 1530	·
6	· 17400	–	· 910	· 62	· 4720	·	70	· 1540	· 1440	·
7	· 13000	–	· 990	· 30	· 4750	·	200	· 2730	· 2590	·
8	· 32500	–	·	· 7	· 5460	·	260	· 5670	–	·
9	· 32200	–	·	· 3–6	· 19 –	6150 –	550	· 6680	–	·
10	· 50500	–	·	· 2–4	· 18 –	6310	· 1770	· 7080	–	·

*Only jump in molar heat capacity [accord. 25, 26]
Points beneath the phase symbols indicate the existence of the phase concerned. The absence of a phase is indicated by strokes

than those of the other phase transitions. Furthermore, transitions either between smectic phases with order within the layers or between smectic phases without order within the layers (see Section 1.4.1.1) exhibit small (S_I-S_H; S_G-S_F; S_C-S_A), extremely small (S_G-S_F; S_F-S_I) or even vanishing values (S_C-S_A) of the transition enthalpy. Transitions between smectic phases with and without order within the layers (S_E-S_C; S_F-S_C; S_I-S_C), also transitions between the liquid-crystalline phases (N, S) and the isotropic liquid show significantly higher transition enthalpies which are obviously due to stronger structural differences between these phases. We can generalize that transitions with strong structural changes exhibit high transition enthalpies, and phase transitions with small structural changes exhibit relatively low transitions enthalpies. In this connection it should be noted that the same statement is also applicable to the volume changes at the phase transitions.

In many cases the transition crystalline → liquid-crystalline can be supercooled, i.e., the liquid crystalline phase may exist as metastable phase below the melting temperature. For the transition isotropic → liquid-crystalline as well as for transitions between several liquid-crystalline phases supercooling (or over-heating) effects are, in general, not observed. If the liquid crystalline phases are subjected to an external pressure the transition temperatures are shifted to higher values according to the equation of Clausius and Clapeyron:

$$\frac{dT}{dp} = T \cdot \frac{\Delta V}{\Delta H}, \qquad 2.1$$

where ΔV is the volume change and ΔH the change of enthalpy at the phases transition. Generally, the T-p-curves of different transitions have a different slope so that in some cases they can intersect at higher pressures. In this case a triple point occurs where one of the phases disappears. For example, the lower steepness of the

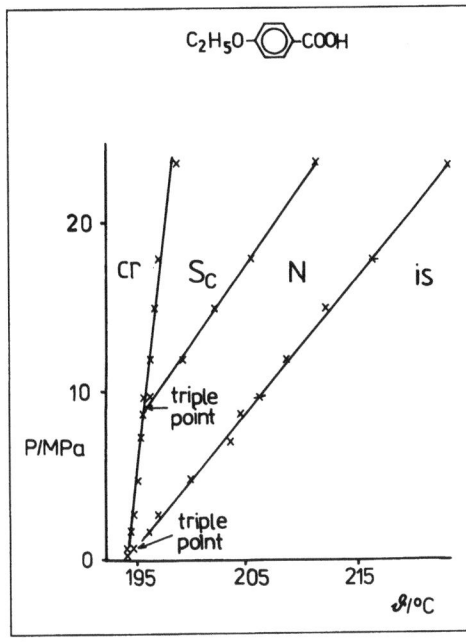

Fig. 2.1. Pressure p vs. temperature ϑ phase diagram of ethoxybenzoic acid [23].

melting curve compared with other transition curves can lead to a pressure-induced stabilization of liquid crystalline phases which are metastable at normal pressure (see Fig. 2.1).

With respect to the large variety of phase transitions the matter of determining their nature (i.e., first or second order) arises. The experimental test to answer this question can be quite difficult because in some cases the transition effects are extremely low; otherwise, different methods can lead to different conclusions. According to all experimental data the transition nematic → isotropic, smectic → isotropic and crystalline → liquid-crystalline is of first order. Also, the transition $S_C \to N$ seems to be of first order, whereas the $S_C \to S_A$ transition is found to be of second order. But there are experimental results for a few materials which can be interpreted as a first order S_C-S_A transition.

According to the microscopic theory of McMillan, the phase transition S_A-N can be of first as well as of second order depending on the ratio T_{AN}/T_{NI}, where T_{AN} is the transition temperature $S_A \to N$ and T_{NI} the transition temperature N → isotropic. When this ratio < 0.88 the $S_A \to N$ transition should become of second order. Indeed, within homologous series and in special binary mixtures such a change of transition order could be observed experimentally, but the critical ratio T_{AN}/T_{NI} was found to be between 0.94 and 0.99.

The transition between a smectic phase with and without order within the layers, respectively, in general, seems to be of first order.

It is extremely difficult to determine the nature of a phase transition between higher ordered smectic phases because the transition effects are often minuscule.

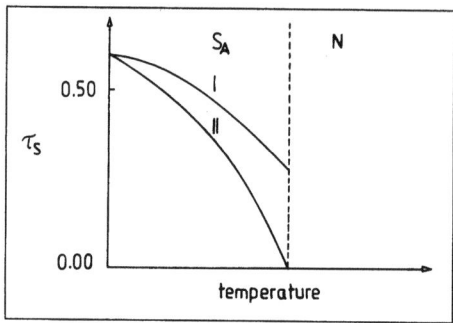

Fig. 2.2. The smectic order parameter τ_s on approaching the phase transition $S_A \to N$. I: phase transition of first order; II: phase transition of second order.

A general phenomenological description of the phase transition in liquid crystals can be performed on the basis of the Landau theory. In this theory the difference between two phases is described by an order parameter. In the ordered phase this order parameter has a finite value, whereas in the non-ordered phase the order parameter is zero. The change of the specific order parameter reflects the order of the corresponding phase transition. For a transition of first order the order parameter vanishes discontinuously, whereas for a transition of second order on approaching the transition temperature the order parameter continuously tends to zero. This is shown in Fig. 2.2 for the smectic order parameter τ_S on approaching the phase transition $S_A \to N$. The order parameter τ_S is a measure of the amplitude of density waves which describes the layer structure (see Section 1.4.1.4).

The general concept of the Landau theory enables a uniform treatment of different phase transitions and gives a qualitative interpretation. For some phase transitions in liquid crystals this concept could be successfully applied (N → isotropic, $S_C \to S_A$, $S_A \to N$), but in some cases, especially for the low-temperature smectic phases, it is very difficult to define suitable order parameters which can be determined experimentally.

The liquid crystalline state is also a suitable object to study multicritical phenomena. A multicritical point is, per definition, a point in a thermodynamic phase plane (temperature-concentration, temperature-pressure) at which three second-order phase boundaries (or two second-order and one first-order phase boundaries) meet. At this point the three phases become indistinguishable-in contrast to a triple point where three first-order phase boundaries intersect and the phases coexist at this point.

The most studied multicritical point in liquid crystals is the so-called NAC-point where the first-order $S_C \to N$ transition curve and the second-order curves $S_C \to S_A$ and $S_A \to N$ of a binary system intersect. It is interesting that, far from the multicritical point, the $S_A \to N$ transition is of first order, but at a definite point near to the multicritical point the $S_A \to N$ transition changes from first to second order. This point is called a tricritical point.

As an experimental example, in Fig. 2.3a the phase diagram of a binary system is shown. Along the $S_C \to S_A$ transition line no transition enthalpy is measured. Otherwise, the entropy of the $S_A \to N$ transition (Fig. 2.3b) decreases and vanishes at

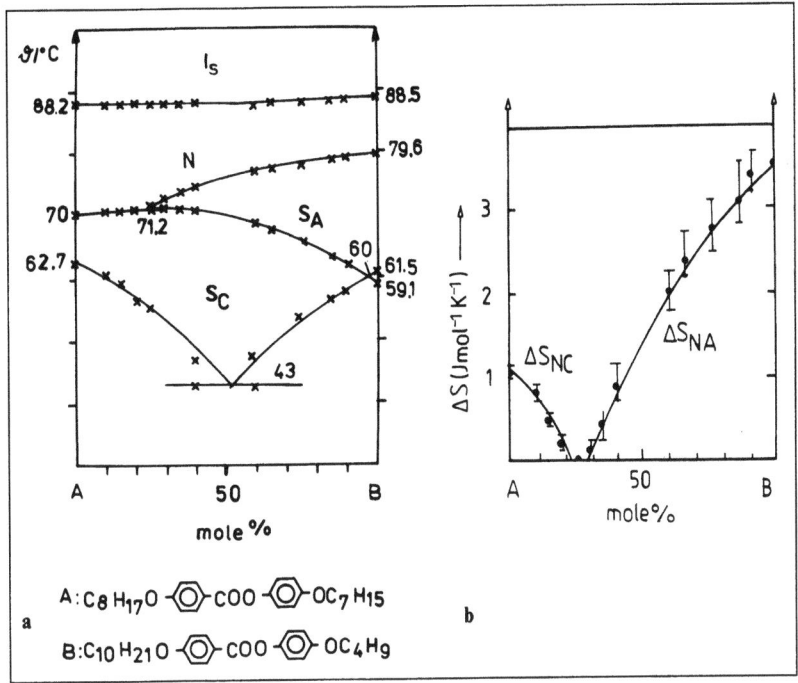

Fig. 2.3. a) Concentration-temperature phase diagram for the binary system with components A and B. b) Entropies ΔS at the phase transitions N → S_C (ΔS_{NC}) and N → S_A (ΔS_{NA}) for the binary mixture with components A and B [24].

a distinct concentration which corresponds to the tricritical point. The entropy at the S_C → N transition also decreases and vanishes at the multicritical point, which is the intersection point of the three transition lines (Fig. 2.3b). Recently, also in the transition curve S_C^* → S_A of a binary mixture, a tricritical point was found.

It should be noted that a tricritical point of the S_A-cholesteric phase boundary line as well as a multicritical NAC point can occur in a pressure-temperature phase diagram of a single liquid crystalline compound.

Comparing the clearing temperatures (transition temperature liquid crystalline → isotropic) of different liquid crystalline substances, it is found that, in general, for structurally similar compounds the clearing temperature increases with increasing length of the molecules. This is illustrated in Table 2.2 where the transition temperatures nematic → isotropic (T_{NI}) are listed for four compounds with similar chemical structure. It is also obvious that by lateral branches the clearing temperature is markedly reduced.

These findings can be interpreted by the generalized Van der Waals theory of the nematic state, according to which there is a correlation between the clearing temperature, the transition entropy ΔS_{NI}, and the volume change ΔV_{NI} at the clearing temperature on the one hand, and the anisotropy of the molecules on the

Table 2.2. Clearing temperature ϑ_{NI} [°C] of some nematic compounds.

Compound	ϑ_{NI}
CH_3O–⟨⟩–$CH{=}N$–⟨⟩–OCH_3	99
CH_3O–⟨⟩–$CH{=}N$–⟨⟩–⟨⟩–OCH_3	318
CH_3O–⟨⟩–$CH{=}N$–⟨⟩–⟨⟩–$N{=}CH$–⟨⟩–OCH_3	> 390*
CH_3O–⟨⟩–$CH{=}N$–⟨⟩–⟨⟩(–$N{=}CH$–⟨⟩–OCH_3)–$N{=}CH$–⟨⟩–OCH_3	218

* decomposition of the substance before clearing

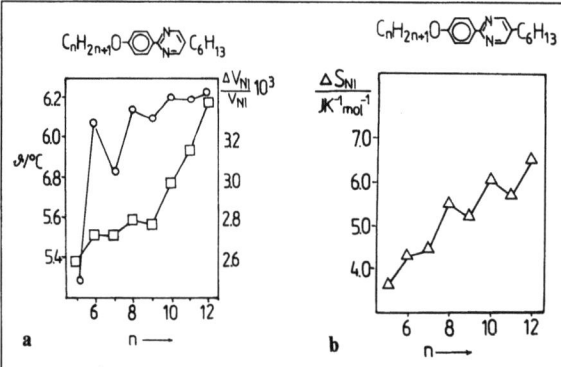

Fig. 2.4. The clearing temperature ϑ_{NI}, the relative volume change $\Delta V_{NI}/V_{NI}$, (a) and the transition entropy ΔS_{NI} (b) plotted against the carbon number n in the alkyloxy chain of homologous 5-n-hexyl-2-(4-n-alkyloxyphenyl)pyrimidines. V_{NI}: molar volume of the nematic phase at ϑ_{NI}; ΔV_{NI}: volume change of the phase transition nematic → isotropic [6].

other hand. In this theory the molecular anisotropy is expressed by the length-to-breadth ratio of the molecules (see Section 1.4.1.4).

The close relation between the molecular anisotropy and T_{NI}, ΔS_{NI}, ΔV_{NI} becomes clear by considering a homologous series in which the neighboring members of the series are distinguished only by one or two CH_2-groups, respectively. In many cases, within a homologous series a pronounced alternation (odd-even effect) of the clearing temperature and also of ΔS_{NI}, ΔV_{NI} and of the orientational order parameter S (see

Section 1.4.1.1) is observed whereby the alternation is most pronounced for the short alkyl chain lengths (see Fig. 2.4).

The Van der Waals theory of nematics is able to give a qualitative interpretation of the alternation effect. With lengthening of the alkyl chain from even to odd numbers of carbon atoms an extra bond is nearly parallel to the molecular long axis. If the chain is lengthened from odd to even numbers of carbon atoms the extra bond is more inclined with respect to the long molecular axis. This conclusion is also applicable if a carbon atom is substituted by a heteroatom such as oxygen in alkyloxy chains.

Because of the alternation of the bonds, also the length-to-breadth ratio of the molecules shows an alternating increase with increasing chain length.

2.2 Optical Anisotropy

In the liquid crystalline state the parallel order of anisotropic molecules (rod-like, disc-shaped molecules) gives rise to the anisotropy of macroscopic physical properties. In this way the anisotropy of the molecular polarizability causes the optical anisotropy of the liquid-crystalline state, i.e., liquid crystals are birefringent like solid crystals. An exception are the cubic liquid crystals which behave optically isotropic. In this connection it should be mentioned that optical anisotropy led to their discovery in 1888 (see Section 1.1).

In this section we primarily consider the optical anisotropy of such liquid crystals which are composed of rod-like molecules.

2.2.1 Optically Uniaxial Liquid Crystals

As follows from the structural symmetry, nematic, S_A, and S_B phases are optically uniaxial whereby the director is identical to the optical axis. In optically uniaxial states there are two principical refractive indices which are related to the ordinary ray (n_o) and to the extraordinary ray (n_e). The optical properties of an anisotropic material can be illustrated by an indicatrix which is, for uniaxial materials, a rotational ellipsoid whereby the rotational axis is identical to the optical axis (Fig. 2.5). The refractive indices and the vibration directions of the ordinary and extraordinary waves which can travel along any direction in the uniaxial liquid crystal are given by the semi-axes of the elliptic sections of the indicatrix which passes to its center and to which the direction at the incidence is normal. For uniaxial materials one semi-axis remains constant. This semi-axis is always the radius of a circular section and this refractive index refers to the ordinary ray. The other axis of the elliptic section which corresponds to n_e depends on the direction of incident light and has an extreme (maximum or minimum) value when the light incidence is perpendicular to the optical axis. For this reason, for the experimental determination of the principal refractive indices the light incidence must be perpendicular to the optical axis. It should be noted that, in the direction of the optical axis a circular

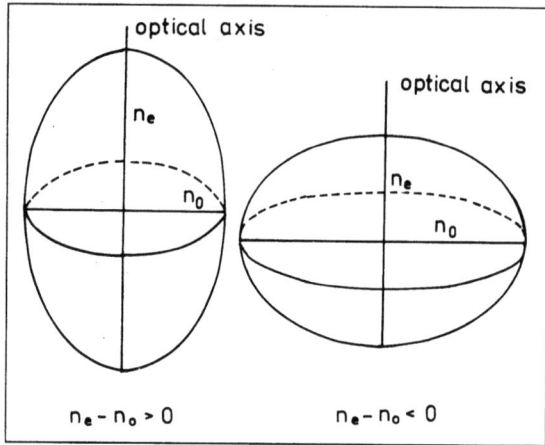

Fig. 2.5. Indicatrix of an optically uniaxial material: a) positive sign of birefringence $(n_e - n_o > 0)$; b) negative sign of birefringence $(n_e - n_o < 0)$.

section results, i.e., in this direction n_e is equal to n_o, the birefringence defined as $n_e - n_o$ is zero.

There are two possibilities for uniaxial materials which are also realized in liquid crystals. In the first case $n_e - n_o > 0$, the birefringence is positive (Fig. 2.5a). In the second case $n_e - n_o < 0$, the birefringence is negative (Fig. 2.5b).

If the principal refractive indices (n_e, n_o) of a uniaxial liquid crystal are correlated with the anisotropy of the polarizability of the molecules it must be taken into consideration that the local field E_{loc} which acts on the molecules within the condensed phase differs from the applied external field E. For liquids the Lorentz field is often used:

$$E_{loc} = \frac{n^2 + 2}{3} E, \qquad 2.2$$

which leads to the Lorenz-Lorentz equation

$$\frac{n^2 - 1}{n^2 + 2} = \frac{\rho N_A}{3 \varepsilon_o M} \bar{\alpha}, \qquad 2.3$$

where ρ = density, $\bar{\alpha}$ = average molecular polarizability, N_A = Avogadro's number; M = molar mass, and $\varepsilon_o = 8.86 \cdot 10^{-12}$ As/Vm.

For liquid crystals the additional problem arises that the local field is anisotropic. There are a few theoretical approximations to describe the local field in liquid crystals. For a more qualitative interpretation of the experimental results it seems to be sufficient to use the equations of Vuks. These equations are relatively often applied in optical studies on liquid crystals (especially for nematics) although the local field in this approximation is isotropic:

$$\frac{n_e^2 - 1}{\overline{n^2} + 2} = \frac{\rho N_A}{3\varepsilon_o M} \bar{\alpha}_\parallel \qquad 2.4$$

$$\frac{n_o^2 - 1}{\overline{n^2} + 2} = \frac{\rho N_A}{3\varepsilon_o M} \bar{\alpha}_\perp, \qquad 2.5$$

with $\overline{n^2} = \frac{1}{3}(n_e^2 + 2n_o^2)$.

The average polarizability parallel and perpendicular to the optical axis can be expressed as

$$\bar{\alpha}_\parallel = \bar{\alpha} + \tfrac{2}{3}(\alpha_1 - \alpha_t) S \qquad 2.6a$$

$$\bar{\alpha}_\perp = \bar{\alpha} - \tfrac{1}{3}(\alpha_1 - \alpha_t) S, \qquad 2.6b$$

where $\bar{\alpha} = \frac{1}{3}(\alpha_1 + 2\alpha_t)$ is the average polarizability and α_1 and α_t are the longitudinal and transversal polarizabilities of the molecules, respectively. From Eqs. 2.4 and 2.5, we obtain

$$\frac{n_e^2 - n_o^2}{\overline{n^2} + 2} = \frac{\rho N_A}{3\varepsilon_o M}(\alpha_1 - \alpha_t) S. \qquad 2.7$$

Because the term on the lefthand side of Eq. 2.7 is a measure of the birefringence, we conclude from Eq. 2.7 that the birefringence of a liquid crystal is determined by the orientational order S, by the anisotropy of the molecular polarizability $(\alpha_1 - \alpha_t)$, and by the reciprocal molar volume ρ/M.

Birefringence of the Nematic Phase

Figure 2.6 shows the characteristic temperature dependence of the principal refractive indices n_e, n_o in the nematic phase of ethyl (4-ethoxybenzylideneamino) α-methylcinnamate for different wavelengths. For comparison, the refractive index n_i of the isotropic liquid is drawn as a function of the temperature. It is seen that the sign of the birefringence is positive $(n_e - n_o > 0)$. With increasing temperature the birefringence decreases; this decrease is particularly pronounced in vicinity of the clearing temperature. At the transition temperature nematic → isotropic the birefringence disappears discontinuously, i.e., at the clearing temperature the birefringence has a finite value, which indicates a phase transition of first order. The dispersion of the refractive indices is normal because they decrease with increasing wavelength. Furthermore, the dispersion decreases in the sequence n_e, n_i, n_o.

The experimental results can be easily understood on the base of Eqs. 2.4 and 2.5. Because for elongated molecules the longitudinal polarizability α_1 is greater than the transverse polarizability α_t and, according to Eqs. 2.4 and 2.5, n_e should be greater than n_o, i.e., the sign of the birefringence is positive.

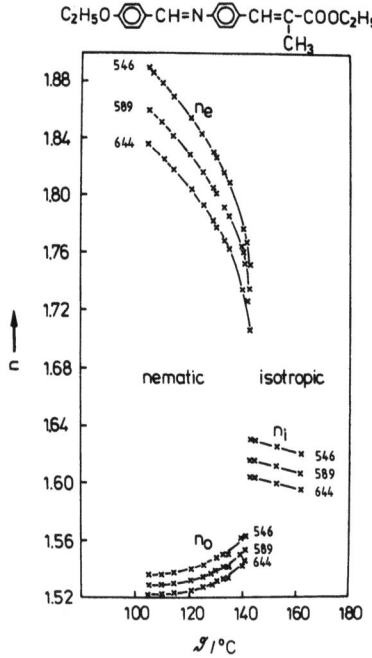

Fig. 2.6. Temperature dependence of the refractive indices in the nematic (n_e, n_o) and isotropic phases (n_i) of ethyl(4-ethoxybenzylidene amino) α-methylcinnamate at the wavelengths 546, 589, and 644 nm [17].

The temperature dependence of n_i is only determined by the temperature change of the density (Eq. 2.3), which explains the linear dependence. The temperature change of the birefringence not only depends on the temperature change of the density, but mainly on the temperature change of the orientational order S (see Eq. 2.7). The strong decrease of $n_e - n_o$ even near the clearing temperature is caused by the drop of S.

The dispersion of the refractive indices in the visible region is largely determined by the absorption in the ultraviolet (u.v.) region. Homogeneously nematic layers exhibit a distinct u.v. dichroism. Generally, the extraordinary waves are much more absorbed than the ordinary waves, whereas the integral absorption of the isotropic liquid has an intermediate value. This behavior is plausible for a parallel alignment of molecules possessing an electron transition moment (which is responsible for the u.v. absorption) parallel to the long molecular axis. According to the theory of dispersion, these differences in absorption lead to the observed differences in the dispersion of the refractive indices.

Within homologous series often an alternation of the birefringence is found (see Fig. 2.7). This alternation can be interpreted by an alternating change of the molecular polarizability anisotropy caused by the alternation of the bonding angle in the aliphatic chains.

In Fig. 2.8 the principal refractive indices (n_e, n_o) for a discotic nematic phase (N_D) are shown in dependence on the temperature. The essential difference from nematic

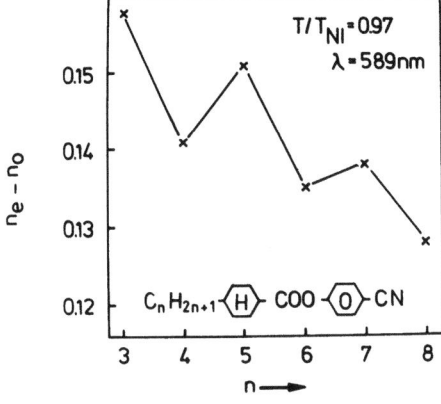

Fig. 2.7. The birefringence of the nematic phase of homologous 4-cyanophenyl 4-n-alkylcyclohexane carboxylates for a constant reduced temperature $T/T_{NI} = 0.97$, where T_{NI} is the clearing temperature. n: number of carbon atoms of the alkyl chain [20].

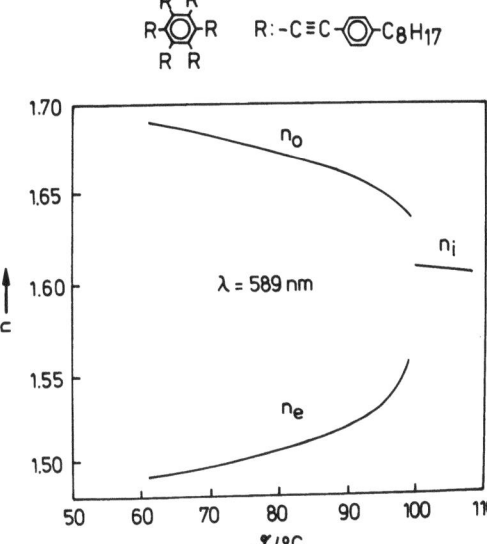

Fig. 2.8. Temperature dependence of the refractive indices in the discotic nematic phase (n_e, n_o) and the isotropic phase (n_i) of hexakis[(4-n-octylphenyl)ethinyl] benzene. ($\lambda = 589$ nm) [8].

phases composed of rod-like molecules is the negative sign of the birefringence ($n_e - n_o < 0$) which is the result of the structure of the discotic nematic phase. The N_D phase is formed by disc-shaped molecules which are aligned, on average, perpendicular to the optical axis (see Section 1.4.2). In this case n_e is measured for light waves polarized parallel to the optical axis which is perpendicular to the planes of molecules. But perpendicular to the planes of the disc-shaped molecule the polarizability has the smallest value.

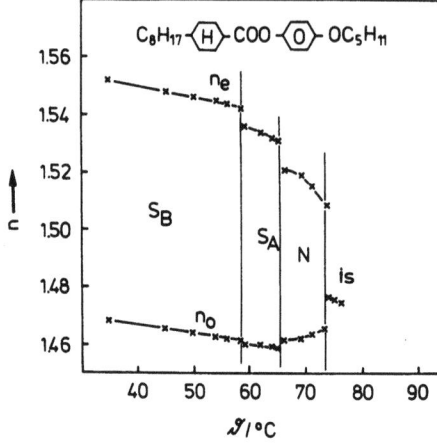

Fig. 2.9. Temperature dependence of n_e and n_o in the nematic, S_A and S_B phase of 4-n-pentyloxyphenyl 4'-n-octylcyclohexane-carboxylate ($\lambda = 589$ nm) [5].

Birefringence of the Smectic Phases S_A and S_B

Figure 2.9 shows the temperature dependence of the refractive indices for a material which possesses, in addition to the nematic phase, the uniaxial smectic phases S_A and S_B. It is seen that, in all liquid crystalline phases, the birefringence decreases with increasing temperature. Otherwise, the temperature dependence clearly increases in the series S_B, S_A, N. At the phase transitions $N \to S_A$ and $S_A \to S_B$ a jump of the birefringence occurs.

The difference in the temperature dependence of the birefringence is obviously the result of the different temperature dependence of the orientational order S. The discontinuities of the birefringence at the phase transitions can be interpreted by the discontinuous change of the orientational order, thus indicating phase transitions of first order. But there are substances for which a continuous change of the birefringence is found at the phase transition $N \to S_A$. This points to a phase transition of second order.

Birefringence of Cholesteric Liquid Crystals

In cholesteric phases (see Section 1.4.1.1) the helix axis is the optical axis of the uniaxial structure. The extraordinary waves vibrate parallel to the optical axis and, therefore, perpendicular to the long molecular axes. Perpendicular to the helix axis (this is the vibration direction of the ordinary waves) all directions of the long molecular axes are realized. Because for rod-like molecules $\alpha_1 - \alpha_t > 0$ it is reasonable that $n_e < n_o$ and the sign of the birefringence is negative, in contrast to nematic phases. This is shown for the cholesteric phase of cholesterol decanoate (Fig. 2.10). At the transition into the S_A phase the birefringence is nearly doubled and the sign of birefringence changes from negative to positive.

According to the theory, the refractive indices of the cholesteric state can be correlated with the principal refractive indices of the non-twisted nematic state $n(N)$

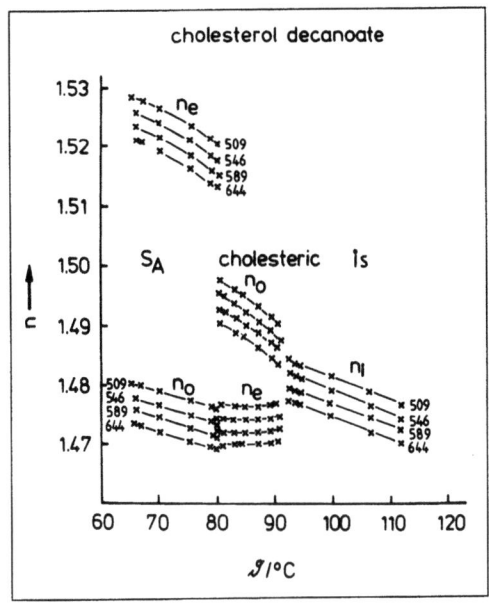

Fig. 2.10. Temperature dependence of the refractive indices in the cholesteric, S_A (n_e, n_o) and isotropic phase (n_i) of cholesterol decanoate at the wavelengths 509, 546, 589, and 644 nm [18].

(formally obtained by the unwinding of the cholesteric structure):

$$n_e = n_o(N) \qquad (2.8a)$$

$$n_o = \sqrt{\tfrac{1}{2}(n_e^2(N) + n_o^2(N))}. \qquad (2.8b)$$

For racemic mixtures of cholesteric enantiomeric substances which behave nematically these relations can be verified experimentally. Because the arrangement of the molecules in the nematic phase is not too different from that of the S_A phase, it is plausible that n_o at the S_A phase is the continuation of n_e in the cholesteric phase, and n_o of the cholesteric phase is about the average value $\tfrac{1}{2}(n_e + n_o)$ of the S_A phase.

2.2.2 Optically Biaxial Liquid Crystals

Smectic C Phases

The S_C phase is distinguished from the S_A phase by the tilt of the director with respect to the layer normal. The structure of the S_C phase exhibits monoclinic symmetry. The point group C_{2h} contains only a two-fold rotation axis parallel to the layers and normal to the long molecular axis, a reflection plane normal to the two-fold axis, and a center of inversion. As a consequence of the reduced symmetry the orientational order is, in general, no longer strictly uniaxial if the elongated molecules themselves have a anisotropy in the plane perpendicular to the long molecular axis. The molecules will preferentially orient with their short axes either parallel or perpendicular to the two-fold symmetry axis. In addition, the fluctuations of the

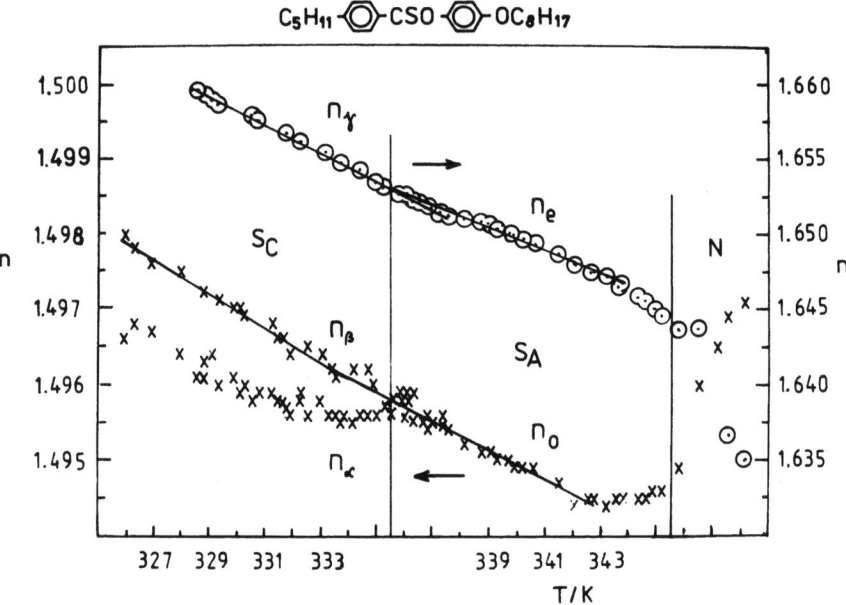

Fig. 2.11. Temperature dependence of the refractive indices in the nematic, S_A (n_e, n_0) and S_C phase (n_α, n_β, n_γ) of 4-n-pentylphenylthio-4-n-nonyloxybenzoate ($\lambda = 589$ nm) [13].

molecules around their preferred direction no longer have rotational symmetry. For this reason the S_C phase behaves biaxially, which can be easily detected by conoscopic investigations.

Up to now, only for a few materials have the principal refractive indices been measured. Figure 2.11 presents the principal refractive indices for the S_A and S_C phase of 4-n-pentylphenylthio 4-n-nonyloxybenzoate.

In the biaxial S_C phase the refractive indices are designated as n_α, n_β, n_γ where the subscripts indicate the main axes of the indicatrix (Fig. 2.12). One of the principal axes (n_γ) coincides with the direction of the director.

The axis n_α of the indicatrix is perpendicular to the director and lies in the tilt plane. The third axis n_β is along the normal of the tilt plane and parallel to the layers (Fig. 2.12). It is seen from Fig. 2.11 that at the phase transition n_e evolves into n_γ, whereas n_o splits into n_α and n_β. The sign of the biaxiality is positive and the angle between the optical axes is relatively low (5° – 14°), depending on the temperature.

Also, the higher ordered tilted smectic phases (S_I, S_F, S_G, S_H....) and the S_E phase (which exhibit an orthorhombic unit cell) are optically biaxial. But, up to now, measurements of the birefringence have been impossible because of the difficulties in obtaining homogeneously oriented monodomains.

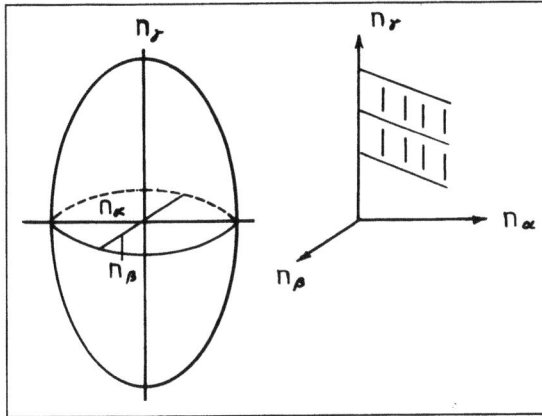

Fig. 2.12. The indicatrix and the principal refractive indices $(n_\gamma, n_\beta, n_\alpha)$ of the S_C phase related to its structure.

It should be mentioned that, for some thermotropic liquid crystals, biaxial nematic phases could be detected. The biaxiality was found to be weak, and the angle between the optical axes is in the same order of magnitude as for S_C phases.

2.3 Optical Properties of Twisted Liquid Crystalline Phases

2.3.1 Cholesteric Phases

The helix structure of the cholesteric phase (see Section 1.4.1.1.) gives rise to some remarkable optical properties. For the case $p = \lambda$ or $p \gg \lambda$ (p = pitch of the cholesteric structure) on proceeding in direction of the helix axis the principal polarizabilities perpendicular to the axis rotate continuously. Therefore, in a narrow wavelength range around λ_o a selective reflection of circularly polarized light takes place. If a light beam is incident parallel to the helix axis it is split into its two circularly polarized components. Depending on the left or right handedness of the helix, one circularly polarized component is transmitted, whereas the other one is totally reflected.

The d-helix reflects right circularly polarized light, whereas left circularly polarized light is transmitted. Inversely, with a l-helix left circularly polarized light is reflected and right circularly polarized light is transmitted. According to the theory of de Vries there is a simple relation between the maximum reflection wavelength λ_o and the pitch p of the helix

$$\lambda_o = \bar{n} \cdot p \,, \qquad 2.9$$

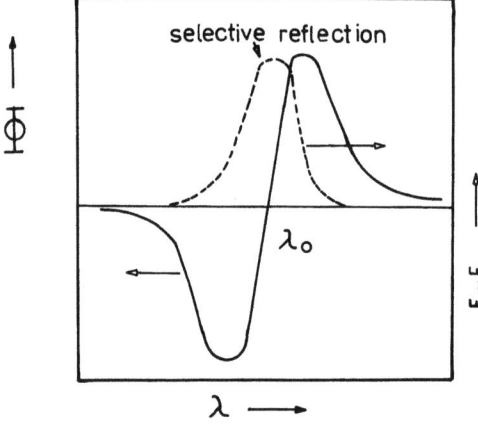

Fig. 2.13. The rotatory power Φ and the selective reflection $E_l - E_r$, respectively, of a cholesteric phase as function of the wavelength (schematic representation).

where $\bar{n}$ is an average refractive index $\bar{n} = \frac{1}{2}(n_e + n_o)$. The width $\Delta\lambda$ of the reflection band is given by

$$\Delta\lambda = \Delta n \cdot p , \qquad 2.10$$

where $\Delta n = n_e - n_o$ is related to two-dimensional quasi-nematic layers.

If the reflection wavelength λ_o is in the visible region, brilliant reflection colors are visible.

In cholesteric materials with an additional smectic phase on approaching the transition cholesteric → smectic the pitch and, hence, the reflection color are strongly temperature dependent. Thin, homogeneously aligned cholesteric layers of such compounds can therefore be used to directly display the temperature distribution of a surface by the distribution of the reflection colors (thermotopography).

In the selective reflection region (see Fig. 2.13) the cholesteric sample exhibits a quasi-circular dichroism which is caused by reflection of light and which must be clearly distinguished from the usual circular dichroism that is due to the light absorption of chiral molecules.

For obliquely incident light the reflection band is shifted to shorter wavelengths and higher order reflection bands occur which are absent at normal incidence.

The dependence of the reflection wavelength λ_o on the angle of incidence ϕ can be described by a formula which is analogous to Bragg's equation:

$$\lambda_o = 2d\bar{n} \sin\phi , \qquad 2.11$$

with $d = p/2$.

Another striking property of cholesteric liquid crystals is the rotation of the plane of polarized light. The rotation power can amount to several thousands of degrees per mm, which is higher by several orders of magnitude than that of optically active crystals. According to the theory of de Vries the rotation per sample thickness Φ can be expressed by

$$\Phi = \frac{-2\pi \alpha^2}{p \cdot 8 \lambda'(1 - \lambda'^2)}, \qquad 2.12$$

where

$$\alpha = \frac{n_e - n_o}{\frac{1}{2}(n_e + n_o)}$$

$$\lambda' = \lambda/\lambda_o .$$

For the borderline case $\lambda' \ll 1$ Eq. 2.12 is reduced to:

$$\Phi = -\frac{2\pi}{p} \frac{\alpha^2}{8\lambda'^2} . \qquad 2.13$$

In this case the light waves are nearly linearly polarized.

In the region of reflection the rotation dispersion is anomalous and the sign of the rotation is opposite on opposite sides of the reflection band. Contrary to the Cotton effect of chiral molecules, this wavelength region is not separated by an absorption band, but by a reflection band (see Fig. 2.13).

It is interesting that, by addition of a small amount of chiral (also non-liquid-crystalline) materials to a nematic matrix a cholesteric helix structure can be induced. Obviously, the symmetry properties of chiral molecules are transformed to the whole liquid crystalline system by the intermolecular interaction.

Enantiomeric dopands twist a nematic matrix in an inverse sense. In racemic mixtures of cholesteric compounds the helix of the enantiomers is compensated so that a nematic structure with $p \to \infty$ results.

In many cases, and with not too high concentrations of the chiral dopant, the pitch length is inversely proportional to the mole fraction x_G of the chiral guest compound:

$$p^{-1} = P \cdot x_G \qquad T = \text{const.} \qquad 2.14$$

The proportional factor P is called "helical twisting power" (HTP) and is a measure of the twisting property of a chiral dopant with respect to a given nematic matrix.

2.3.2 Chiral Smectic C Phases (S_C^*)

In S_C phases composed of chiral molecules as a consequence of the molecular chirality the molecular director exhibits a helical structure, the pitch of which corresponds to a complete rotation of the director through an angle of 2π. The period along the helix axis is identical with p-in contrast to cholesteric liquids in which the period corresponds to $p/2$. This twisted S_C phase is designated as S_C^* phase.

For $\lambda \gg p$ the optical properties of the S_C^* phase are quite similar to that of cholesteric phases. Like cholesteric phases S_C^* phases behave optically uniaxial, whereby the optical axis is identical to the helix axis. S_C^* phases show selective reflection, circular dichroism, and rotatory power which can be described by analogous formulas. In contrast to cholesteric phases in S_C^* phases an additional Bragg reflection band occurs at obliquely incident light.

2.4 Magnetic Properties

Like most of the organic substances liquid crystals are usually diamagnetic[1], i.e., the magnetic susceptibility is small and negative ($\chi < 0$). Because of the anisotropy of the molecular magnetic susceptibility the diamagnetic susceptibility is also anisotropic. This is demonstrated for the nematic phase of 4(4-n-heptylcyclohexyl)benzonitrile (Fig. 2.14).

It is seen that the mass diamagnetic susceptibility parallel to the director ($\chi_\parallel$) is greater than that perpendicular to the director ($\chi_\perp$); the sign of the diamagnetic anisotropy is positive ($\Delta\chi = \chi_\parallel - \chi_\perp > 0$). The anisotropy decreases with increasing temperature. For substances with an additional S_A phase at the transition $N \rightarrow S_A$ a jump of $\Delta\chi$ is generally observed (see Fig. 2.15). Because the diamagnetic moment of the molecules is very small, the interaction between the magnetic moments can be ignored so that the magnetic field acting on a molecule can be taken as equal to the external field. For this reason the principal susceptibilities of the liquid crystalline phases ($\chi_\parallel$, $\chi_\perp$) can be directly related to the molar magnetic polarizabilities (γ_1, γ_t) taking into consideration the imperfect molecular order by the orientational order S:

$$\chi_\parallel = \frac{\rho}{M} \cdot N_A \left[\bar{\gamma} + \frac{2}{3} S(\gamma_1 - \gamma_t)\right] \qquad 2.15$$

$$\chi_\perp = \frac{\rho}{M} \cdot N_A \left[\bar{\gamma} - \frac{1}{3} S(\gamma_1 - \gamma_t)\right], \qquad 2.16$$

where γ_1 is the longitudinal and γ_t the transversal magnetic polarizability of the molecules, and ρ = density, $\gamma_1 - \gamma_t$ = anisotropy of the molecular magnetic polarizability, M = molar mass, N_A = Avogadro's number, $\bar{\gamma} = (\gamma_1 + 2\gamma_t)/3$ = average magnetic polarizability, and S = orientational order.

Combination of Eqs. 2.15 and 2.16 leads to

$$\Delta\chi = \chi_\parallel - \chi_\perp = \frac{\rho}{M} \cdot N_A \ (\gamma_1 - \gamma_t) \ S = S(\chi_\parallel - \chi_\perp)_{s=1}. \qquad 2.17$$

It follows from Eq. 2.17 that the temperature dependence of the anisotropy of the diamagnetic susceptibility is mainly determined by the temperature change of S.

---

[1] There are a few liquid crystals with paramagnetic properties, e.g., free radicals or complexes with paramagnetic metal ligands.

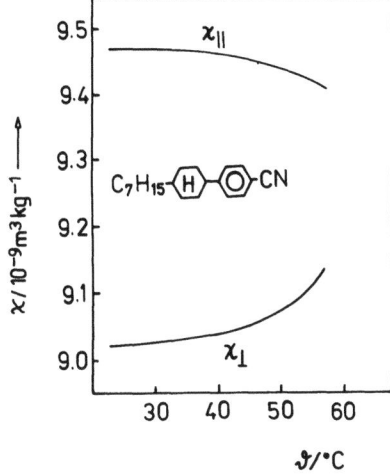

Fig. 2.14. Temperature dependence of the diamagnetic mass susceptibilities parallel ($\chi_\parallel$) and perpendicular ($\chi_\perp$) to the director in the nematic phase of 4-(trans-4-n-heptylcyclohexyl) benzonitrile [21].

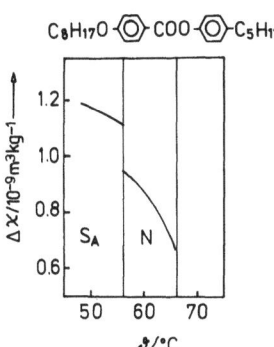

Fig. 2.15. The anisotropy of diamagnetic mass susceptibility as function of the temperature in the nematic and S_A phase of 4-n-pentylphenyl 4'-n-octyloxy-benzoate [3].

A discontinuity of $\Delta\chi$ at the phase transition $N \rightarrow S_A$ (Fig. 2.15) for 4-n-pentylphenyl 4-n-octyloxy-benzoate is due to a discontinuous increase of S which points to a phase transition of first order.

Eq. 2.17 can be used to calculate the orientational order S from measurements of the anisotropy of the diamagnetic susceptibility. In this case $(\chi_\parallel - \chi_\perp)_{s=1}$ must be determined from the susceptibilities of a perfectly ordered monocrystal.

Generally, for rod-like molecules $\gamma_l - \gamma_t$ and, therefore, $\chi_\parallel - \chi_\perp$ are positive (see Fig. 2.14). The positive sign of $(\gamma_l - \gamma_t)$ can be interpreted in the following way. The delocalized π-electrons of the benzene ring form a "ring current". If the magnetic field is perpendicular to the plane of the ring the counteracting induced moments should be stronger than for the field direction along the ring. From this follows a positive anisotropy of the susceptibility which is proportional to the number of the rings.

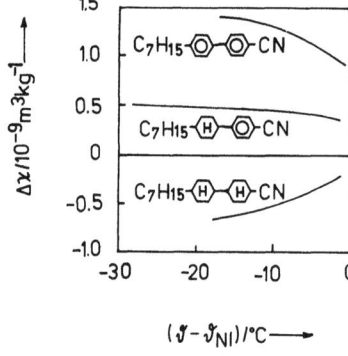

Fig. 2.16. Anisotropy of the diamagnetic mass susceptibility vs. temperature difference $\vartheta - \vartheta_{NI}$ in the nematic phase of three compounds which are distinguished by the number of benzene rings [10].

If benzene rings are substituted by cyclohexane rings the anisotropy of susceptibility drops. For non-aromatic liquid crystals the sign of $\Delta\chi$ is found to be negative. This is illustrated in Fig. 2.16 where $\Delta\chi$ is presented for three analogous nematic materials.

The anisotropy of the diamagnetic susceptibility plays an important role for the alignment of nematic liquid crystals by means of magnetic fields. For sufficiently high strengths of the magnetic field the director of a nematic phase with $\Delta\chi > 0$ can be oriented in field direction by the coupling of the anisotropic part of the external magnetic field. The threshold H_o of this reorientation is given by

$$H_o = \frac{\pi}{d} \sqrt{k_{11}/\mu_o \cdot \Delta\chi} \,, \qquad 2.18$$

where d is the sample thickness and k_{11} is the elastic constant of splay deformation (Section 2.6.1). To achieve a well-ordered nematic monocrystal characterized by a uniform alignment of the molecular director magnetic fields of $\geqslant 0.5$ Tesla are necessary. It is important that this uniform alignment can often be preserved on cooling into smectic phases, whereby it is advantageous if the orienting magnetic field acts during the phase transition. Liquid crystalline samples oriented in this way are used for X-ray investigations or for measurements of the anisotropy of the dielectric permittivity or the viscosity.

2.5 Dielectric Properties

2.5.1 Static and Low-Frequency Dielectric Anisotropy

The static dielectric permittivity ε of an isotropic non-associated liquid can be approximately expressed by Onsagér's equation:

$$\varepsilon_0(\varepsilon - 1) = N \cdot F \cdot h \left(\alpha + \frac{\mu^2}{3kT} F \right), \qquad 2.19$$

where α is the molecular polarizability, μ is the dipole moment of the molecule, and N is the number of molecules per unit volume. The reaction field factor F and the cavity factor h are parameters of the Onsagér theory, and $F = 1/(1 - f'\alpha)$ with $f' = (3\varepsilon - 2)/(2\varepsilon + 1) \, (1/\varepsilon_0 a^3)$, $h = 3\varepsilon/(2\varepsilon + 1)$, $\varepsilon_0 = 8.86\cdot 10^{-12}$ As/Vm, $k =$ Boltzmann constant, and $a =$ radius of the spherical cavity.

In liquid crystals the dielectric properties are anisotropic. The anisotropy of the dielectric permittivity is of great interest since most of practically relevant electro-optical effects are based on the dielectric anisotropy.

In uniaxial liquid crystals like nematic phases the anisotropy of the dielectric permittivity is defined as $\Delta\varepsilon = \varepsilon_\parallel - \varepsilon_\perp$, where $\varepsilon_\parallel$ and $\varepsilon_\perp$ are the dielectric permittivities parallel and perpendicular to the director, respectively.

Magnitude and sign of the dielectric anisotropy are determined by the molecular structure of the liquid crystals. A general theoretical interpretation of the dielectric anisotropy as a function of the molecular parameters is possible by the Maier–Meier theory which is an extension of the Onsagér theory of polar liquids to nematic liquid crystals. This theory considers the molecules, for simplicity, as spherical, but their polarizability is anisotropic. The point dipole μ in the center of the molecule encloses an angle β with the long molecular axis. It is assumed that $\Delta\varepsilon \ll \bar{\varepsilon}$, where $\bar{\varepsilon}$ is the average dielectric permittivity $\bar{\varepsilon} = (\varepsilon_\parallel + 2\varepsilon_\perp)/3$. This assumption means that the surroundings of the molecules are considered as a continuum.

Furthermore, the nematic order is not influenced by the external electric field. On the basis of these assumptions the following equations for the principal dielectric permittivities $\varepsilon_\parallel$ and $\varepsilon_\perp$ of a nematic "monocrystal" are obtained:

$$(\varepsilon_\parallel - 1)\varepsilon_0 = NFh\left\{\bar{\alpha} + \frac{2}{3}\Delta\alpha S + \frac{F\mu^2}{3kT}[1 - (1 - 3\cos^2\beta)S]\right\} \qquad 2.20$$

$$(\varepsilon_\perp - 1)\varepsilon_0 = NFh\left\{\bar{\alpha} - \frac{1}{3}\Delta\alpha S + \frac{F\mu^2}{3kT}[1 + \frac{1}{2}(1 - 3\cos^2\beta)S]\right\} \qquad 2.21$$

The anisotropy of the dielectric permittivity is given by

$$(\varepsilon_\parallel - \varepsilon_\perp)\varepsilon_0 = NFhS\left[\Delta\alpha - \frac{F\cdot\mu^2}{2kT}(1 - 3\cos^2\beta)\right], \qquad 2.22$$

where $\Delta\alpha = \alpha_l - \alpha_t =$ anisotropy of the molecular polarizability, and α_l, α_t are the polarizabilities parallel and perpendicular to the long molecular axis; $\bar{\alpha} = (\alpha_l + 2\alpha_t)/3 =$ average molecular polarizability and $\beta =$ angle between the total dipole moment μ and the long molecular axis.

The first term in Eq. 2.22 corresponds to the anisotropy of the induced polarization which arises mainly from the displacement of electrons by the action of the external electric field. The second term describes the anisotropy of the orientation polarization which is connected with the orientation of the permanent molecular dipoles in the direction of the field by considering the respective orientational order. In nematic liquid crystals formed by rod-like molecules the polarizability anisotropy is positive ($\Delta\alpha = \alpha_l - \alpha_t > 0$). Therefore, the contribution of the induced polarization to $\Delta\varepsilon$ is also positive, i.e. means, nonpolar nematic compounds ($\mu = 0$) exhibit positive

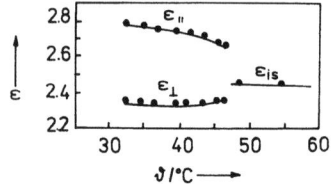

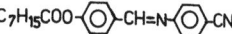

Fig. 2.17. Temperature dependence of the dielectric permittivities parallel ($\varepsilon_\|$) and perpendicular ($\varepsilon_\perp$) to the director in the nematic phase of 4,4'-di-n-heptylazobenzene. ε_{is} is the dielectric permittivity of the isotropic liquid [9].

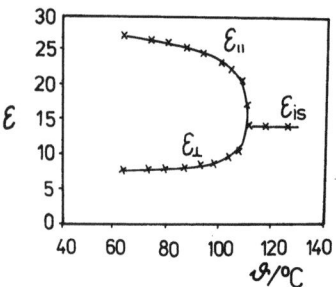

Fig. 2.18. Temperature dependence of the dielectric permittivities in the nematic ($\varepsilon_\|$, $\varepsilon_\perp$) and isotropic phases (ε_{is}) of 4-n-heptylcarbonyloxybenzylidene-4'-cyanoaniline [22].

dielectric anisotropy ($\Delta\varepsilon > 0$). This is shown in Fig. 2.17 where the temperature dependence of $\varepsilon_\|$ and $\varepsilon_\perp$ is presented for the nematic phase of the nonpolar 4,4-di-n-heptylazobenzene. For comparison, the dielectric permittivity of the isotropic liquid (ε_{is}) is shown. Because $\varepsilon = n^2$ for long wavelengths, it is reasonable that the static dielectric anisotropy of non-polar compounds has values similar to those for $n_e^2 - n_o^2$.

The contribution of the orientation polarization depends on the magnitude of the total dipole moment μ and the angle β between μ and the molecular long axis. If $3\cos^2\beta = 1$, which corresponds to the "magic angle" $\beta = 54.74°$, the dipole moment contributes equally to $\varepsilon_\|$ and $\varepsilon_\perp$. For $\beta < 54.74°$ the contribution of the orientation polarization to $\Delta\varepsilon$ is positive, for $\beta > 54.74°$ this contribution is negative. From this, it follows that the sign of the total dielectric anisotropy is determined by the relative magnitude of the terms of the induced polarization and of the orientation polarization.

A strong dipole moment in the direction of the molecular long axis ($\beta = 0$) leads to high values of $\Delta\varepsilon$, i.e., p-cyano-substituted substances (see Fig. 2.18). If the negative contribution of orientation polarization predominates for materials with lateral dipole moments, the dielectric anisotropy becomes negative ($\Delta\varepsilon < 0$). If both contributions are of the same order of magnitude, $\Delta\varepsilon$ is close to zero.

According to Eq. 2.22 the temperature dependence of $\Delta\varepsilon$ is determined by the temperature change of S (for the induced polarization) and by S/T, respectively (for the orientation polarization).

In principle, the Maier–Meier model can also be applied to cholesteric liquid crystals. The dielectric constants parallel ($\varepsilon_{\|h}$) and perpendicular to the helix axis ($\varepsilon_{\perp h}$) are related to the main dielectric constants $\varepsilon_\|$ and $\varepsilon_\perp$ of the untwisted nematic state:

$$\varepsilon_{\|h} = \varepsilon_\perp; \quad \varepsilon_{\perp h} = \tfrac{1}{2}(\varepsilon_\| + \varepsilon_\perp). \qquad 2.23$$

Smectic phases with a tilt of the molecules (S_C, S_F, S_I...) behave biaxially, i.e., the dielectric tensor has three non-equivalent axes. But, until now, for only one substance could the biaxiality of the S_C phase be measured, because it is very difficult to obtain well-oriented monodomains. It was found (analogous to the optical anisotropy) that the biaxiality is very weak, therefore, in most theoretical considerations S_C phases are treated as approximately uniaxial.

2.5.2 Frequency Dependence of the Dielectric Permittivity

If the applied electric field is removed, the reorientation of the induced polarization is faster than 10^{-11} s because of the high mobility of the electrons. Otherwise, the decay of the orientation polarization (which is connected with motion of parts of molecules or even of the whole molecule) takes place more slowly. After removing the electric field the orientation polarization decays with the relaxation time τ because the reorientation of the permanent dipole moments requires a definite time. The relaxation time τ is defined as that time at which the orientation polarization decreases to $1/e$ of the original value.

If the frequency of the applied electric field is not too high the dipoles (and the orientation polarization) can follow the driving field without delay so that the total polarization (including induced and orientation polarization) is measured and corresponds to the static dielectric permittivity $\varepsilon_{(0)}$. At frequencies of $(2'\pi\tau)^{-1}$ the relaxation leads to a noticeable delay between the average polarization of the dipole moments and the applied field. At much higher frequencies the orientation polarization cannot follow the variation of the field so that the residual permittivity $\varepsilon_{(\infty)}$ is only due to the induced polarization.

A typical frequency dependence of the dielectric permittivity of a polar liquid is shown in Fig. 2.19. It is obvious that ε drops in the frequency range of $(2\pi\tau)^{-1}$. Such a dispersion step of ε is absent in nonpolar liquids. In a general theory of the linear response of a dielectric medium in an external electric field the phase lag between the displacement $D = \varepsilon^* \cdot E$ and the external field E in the dispersion region can be expressed by the complex dielectric permittivity ε^*. The frequency dependence of ε^* follows Debye's equation

$$\varepsilon^*_{(\omega)} - \varepsilon_{(\infty)} = \frac{\varepsilon_{(0)} - \varepsilon_{(\infty)}}{1 - i\omega\tau}, \qquad 2.24$$

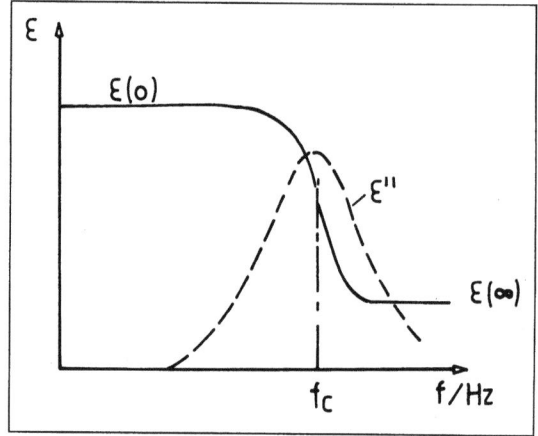

Fig. 2.19. Frequency dependence of the dielectric permittivity ε and the dielectric losses ε'' of a polar liquid (schematic representation).

where the real and imaginary parts of the complex dielectric permittivity $\varepsilon^* = \varepsilon' + i\varepsilon''$ are given by

$$\varepsilon' = \varepsilon_{(\infty)} + \frac{\varepsilon_{(o)} - \varepsilon_{(\infty)}}{1 + \omega^2 \tau^2}, \qquad 2.25$$

$$\varepsilon'' = \frac{\varepsilon_{(o)} - \varepsilon_{(\infty)}}{1 + \omega^2 \tau^2}. \qquad 2.26$$

Equation 2.25 describes the frequency dependence of the real component of ε^*. The dotted curve in Fig. 2.19 corresponds to the frequency dependence of the imaginary part of ε^* which is given by Eq. 2.26. The ratio ε'' to $\varepsilon' - \varepsilon_{(\infty)}$ determines the phase angle of the dielectric losses:

$$\tan \phi = \varepsilon''/[\varepsilon' - \varepsilon_{(\infty)}] = \omega \tau. \qquad 2.27$$

According to Debye's theory, the relaxation time of dipoles can be expressed by

$$\tau = \eta \cdot a^3 / k T \cdot \varepsilon_0, \qquad 2.28$$

where η is the viscosity, a the radius of the molecules, and k is the Boltzmann constant.

Commonly, the relaxation frequency is in the GHz-range. It follows from Eq. 2.28 that in isotropic liquids the temperature dependence of τ is mainly determined by the temperature dependence of η.

The liquid crystalline state is characterized by a long-range order of anisotropic molecules. If we suppose a rod-like geometry of the molecules, we have to decide between the reorientation process of the dipole moments around the long molecular axis (observed in the GHz-range) and the reorientation around the short molecular

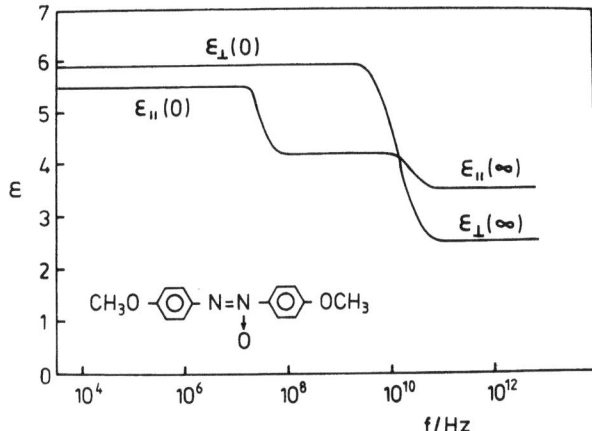

Fig. 2.20. Frequency dependence of the dielectric permittivities $\varepsilon_\|$ and $\varepsilon_\perp$ in the nematic phase of 4,4'-di-methoxyazoxybenzene [14].

axis in the MHz range. Therefore, the anisotropic character of liquid crystals is also reflected in the frequency dependence of the dielectric permittivities $\varepsilon_\|$ and $\varepsilon_\perp$ and is shown for the nematic phase of 4,4'-di-methoxy-azoxybenzene (Fig. 2.20). It is seen that $\varepsilon_\|$ exhibit two dispersion regions, and $\varepsilon_\perp$ (and ε_{is}) exhibits only one. This general behavior can be qualitatively interpreted by the Martin–Meier–Saupe theory which is an extension of Debye's model to nematic liquid crystals by introduction of the nematic potential responsible for the nematic order. According to this model the rotation of the molecules around the long molecular axes is not influenced by the nematic potential. The high-frequency dispersion in $\varepsilon_\perp$ and $\varepsilon_\|$ obviously corresponds to the relaxation of isotropic liquids because the rotation around the long axis is not markedly different in the isotropic and the nematic phase. That also a high-frequency dispersion of $\varepsilon_\|$ can be observed is due to the contribution of the transversal component of the dipole moment to $\varepsilon_\|$ because of the imperfect nematic order.

The low-frequency dispersion of $\varepsilon_\|$ is a special feature of the nematic state. The nematic potential which causes the orientational order of the nematic state hinders the rotation of the molecules around the short axis so that the relaxation time is relatively high. For nematic compounds with $\Delta\varepsilon > 0$ the low-frequency dispersion of $\varepsilon_\|$ can lead to a reversal of the sign of $\Delta\varepsilon$ at a certain frequency. Nematic materials with a relatively low cross-over frequency exhibit a positive dielectric anisotropy ($\Delta\varepsilon > 0$) below, and a negative dielectric anisotropy ($\Delta\varepsilon < 0$) above, and are synthesized for a special electrooptical effect which is based on the two-frequency addressing.

Measurements of the dielectric relaxation give valuable information about the dynamic behavior of various liquid-crystalline phase types. This is illustrated in Fig. 2.21 where the temperature dependence of the relaxation time parallel to the director

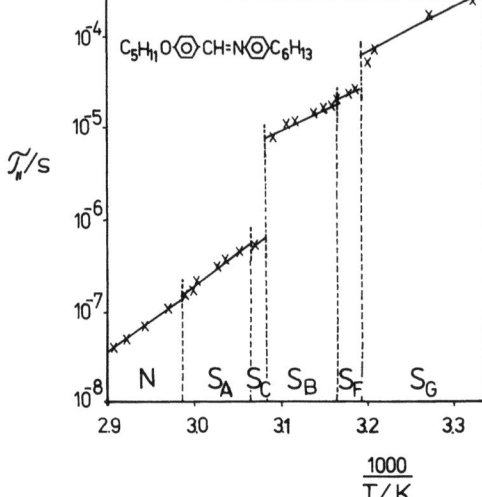

Fig. 2.21. The relaxation time $\tau_\parallel$ in different liquid crystalline phases of 4-n-pentyloxybenzylidene 4'-n-hexylaniline as function of the reciprocal temperature [12].

($\tau_\parallel$) is plotted for a substance exhibiting a rich polymorphism of liquid crystalline phases. From the point of view of molecular dynamics the liquid crystalline phases can be classified into three groups:

A) N, S_A, S_C: The rotation of the molecules (or parts of them) around their long axes is not markedly hindered in comparison to the isotropic liquid. The rotation around the short axis is more or less hindered depending on the degree of order. But there is a continuous change of the relaxation times at the transition $N \to S_A \to S_C$.

B) *Smectic modifications S_I, S_F, S_B(hexatic)*: There is only a small hindrance of the rotation around the long axes, but a stronger hindrance of the rotation around the short molecular axes due to the higher packing density. At the transition from the low ordered S_A phase into the hexatic S_B phase a small jump of $\tau_\parallel$ is observed indicating a discontinuous increase of the hindrance of rotation around the short molecular axis.

C) *Highly-ordered smectic phases, i.e., S_B (crystalline), S_G, S_H, S_K, S_J, S_E*: A stronger hindrance of rotation or else more stochastic movements around the molecular long axes take places. There is a very strong restriction of the rotation around the short molecular axes. Also, at the transition into the highly ordered smectic phases an additional increase of $\tau_\parallel$ is measured, i.e., at the transition S_A or $S_C \to S_B$ (solid-like) or S_B (hexatic) $\to S_E$ $\tau_\parallel$ increases by about one decade. The steps in $\tau_\parallel$ can be correlated to the step-wise increase of the orientational order S.

2.6 Elastic Properties of Liquid Crystals

2.6.1 Elastic Properties of Nematic Liquid Crystals

There are a number of phenomena in liquid crystals which can be described by the treatment of the liquid crystalline state as a continuum, i.e., the elastic deformations caused by external forces such as boundary forces, elastic or magnetic fields. The application of the continuum theory to liquid crystals presumes that the elastic forces should be small compared with intermolecular forces and the characteristic length of the elastic deformations should be large in comparison with the molecular dimensions.

For the nematic state the continuum theory is essentially complete. Like in isotropic liquids, some elastic properties are due to changes in the density which are characterized by suitable elastic constants. But the distinguishing feature of liquid crystals is that the elasticity is related to local variation of the direction of the director, whereas in solid crystals a translational displacement of the molecules gives rise to a deformation. Therefore, in nematic phases there are not restoring elastic forces (as in solids), but rather restoring elastic torques that arise to oppose these deformations.

In nematic liquid crystals the uniform parallel alignment of the local directors represents the equilibrium state which corresponds to the state of minimum free energy. By external forces, i.e., boundary forces, electric or magnetic fields, or material flow, the direction of the local director can be changed. In this way the nematic liquid crystal is deformed whereby the free energy is enhanced. There are three basic types of elastic deformations in nematic liquids to which all deformations can be attributed: splay, bend, and twist deformations (see Fig. 2.22).

The increase of the free energy F per volume relative to its value of the stable uniform orientation can be expressed by three independent elastic constants which are related to the splay (k_{11}), the twist (k_{22}), and the bend (k_{33}):

$$F = \tfrac{1}{2}[k_{11}(\text{div }\vec{n})^2 + k_{22}(\vec{n}\text{ rot }\vec{n})^2 + k_{33}(\vec{n}\times\text{rot }\vec{n})^2], \qquad 2.29$$

where $\vec{n}$ = molecule director.

The free energy density is a quadratic function of the curvature strains where the elastic contants are factors of proportionality. The elastic constants should have the dimension energy per length. The energy of the molecular interaction is estimated to be about 10 kJ mol^{-1} and the molecular dimension is about 2 nm so that for the elastic constants an order of magnitude 10^{-11} N results, which is in agreement with the experimental experience. From this follows that the energy of elastic deformations is by several orders of magnitude smaller than that in solid crystals. A rough estimation shows that the deformation energies are considerably smaller than the average thermic energy. For this reason nematic liquid crystals can be deformed by relatively small external forces (boundary forces, electric or magnetic fields, material flow, temperature gradients).

In the neighborhood of the phase transition nematic $\to S_A$, because of a pretransitional smectic order, a strong divergence of k_{33} and k_{22} is observed,

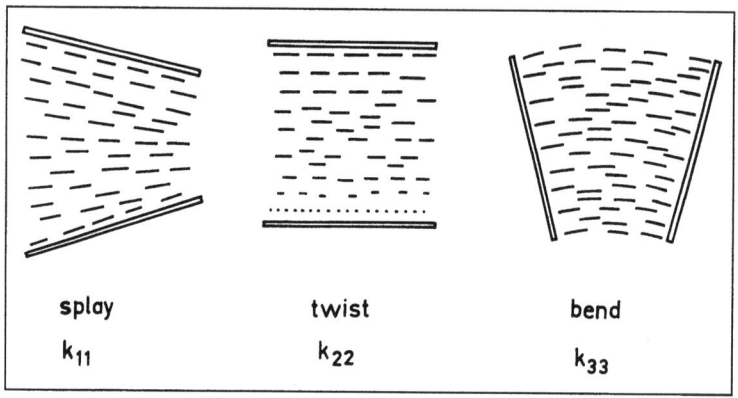

Fig. 2.22. Schematic representation of a a) splay (k_{11}) b) bend (k_{33}), and c) twist (k_{22}) deformation in a nematic liquid crystal.

whereas k_{11} changes continuously. In the vicinity of the transition $N \rightarrow S_C$ all elastic constants diverge. In discotic nematic liquid crystals where the director represents the preferred direction of the short molecular axes, only for a few materials have elastic constants been determined. However, the twist elastic constant k_{22} has not yet been measured, but the values for k_{33} and k_{11} are of the same order of magnitude as for nematics of rod-like molecules.

2.6.2 Uniform Alignment of Nematic Liquid Crystals by Wall Forces

The deformation of nematic liquid crystals by boundaries plays an important role to orient thin samples (2...100 μm). In this case the molecules are uniformly aligned at the substrates by boundary interaction and this fixed orientation can be transferred to the bulk by elastic forces.

There are three basic kinds of uniform orientation. The homeotropic alignment is characterized by an orientation of the director perpendicular to the bounding walls (Fig. 2.23a). In the planar aligned samples the director orientation is uniformly parallel to the substrate (Fig. 2.23b). The intermediate case of orientation is the uniform tilt of the director with respect to the substrate planes (Fig. 2.23c).

From these basic types of orientation some modified cases can be derived in which the director orientation is continuously changed within the nematic sample. Among these orientations the planar twisted layers are of practical importance. In this case the planar oriented director changes its direction to the other wall by $\pi/2$ (or $k \cdot \pi/2$); (see Chapter 6) Uniformly oriented liquid crystalline samples are needed for the measurement of the anisotropy of physical properties and also for the study of electrooptical effects (Section 2.6.3–2.6.5).

There exists a general rule according to which the spontaneous orientation of the nematic liquid crystal between two substrates depends on the relative magnitude of

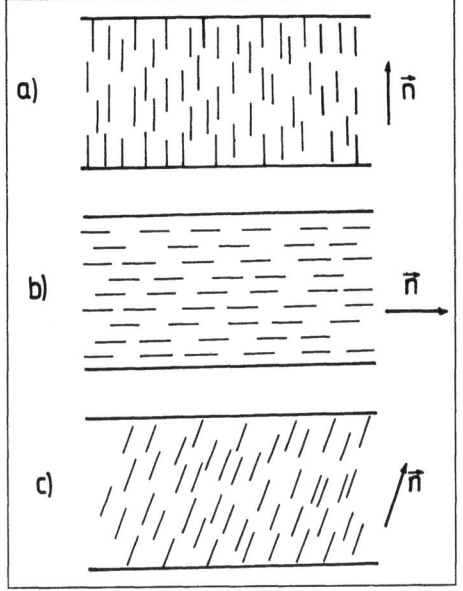

Fig. 2.23. Schematic representation of the a) homeotropic b) planar, and c) inclined alignment of the nematic phase.

the surface energy of the boundary and the surface tension of the liquid crystal. Substrates with a surface energy lower than the surface tension should induce a homeotropic alignment. In the inverse case a planar or also an inclined orientation should be favored. The problem is that it is very difficult to measure these physical properties because small amounts of impurities (also of water) can drastically change these parameters.

Homeotropic alignment

To build up a homeotropic alignment the substrates are coated with a very thin film of special materials such as tensides (hexadecyl trimethylammoniumbromide), lecithin or special metal complexes. In some cases the surface film is generated by direct polymerization of an organosilicon monomer on the substrate. In the sense of the general rule mentioned above, the surface energy of the substrates is reduced by these surfactants. From the point of view of molecular interaction the orientation effect can be interpreted by the attaching of the surfactant molecules on the surface by their polar groups, whereas the long aliphatic chains are aligned perpendicular to the boundary. The aliphatic chains transfer this orientation on the molecules of the nematic bulk.

Planar oriented samples

Planar oriented samples can be achieved by mechanical treatment or chemical modification of the substrates. The often applied method is the rubbing of the substrates in a definite direction by which a microrelief of parallel grooves is generated. The alignment of the elongated molecules of a nematic liquid along the grooves leads to a gain of elastic energy. More uniform microgrooves are obtained by oblique evaporation of some materials such as oxides or metals whereby the angle between the direction of the evaporation and the normal of incidence must be between 45° and 80°.

Furthermore, the chemical nature of the surface can favor the planar alignment. If surfactant molecules are aligned uniformly parallel to the substrates, i.e., by rubbing, they transfer their orientation to the neighboring molecules of the nematic liquid crystal. In most cases, both effects (the pure geometric effect and the chemical nature of the substrate) cooperate and cannot be separated.

Samples with inclined director orientation

Nematic layers with a fixed uniform tilt of the director with respect to the substrate plane can be achieved by oblique evaporation of the substrates with high angles of evaporation (between 80° and 90°). But, also specific chemical treatment of the substrates can generate an inclined director orientation.

It is very important that the uniform alignment of nematic phases can often be transferred to the smectic state by a slow cooling of the nematic state. An alignment of smectic phases by boundary forces is very difficult. An exception is the S_A phase, which can be relatively easily obtained in a homeotropic alignment. Also, the Grandjean texture of cholesteric phases where the helix axis is perpendicular to the substrates often occurs by a small mechanical pressure on the substrates.

2.6.3 Deformation of Nematic Liquid Crystals by an Electric Field

The presence of an electric field leads to extra terms in the expression of the free energy density:

$$-\tfrac{1}{2}\varepsilon_0\,\varepsilon\cdot E^2 - \tfrac{1}{2}\varepsilon_0\,\Delta\varepsilon(\vec{E}\,\vec{n})^2.$$

The first term can be omitted as it is independent of the orientation of the director. As a consequence of the dielectric anisotropy (considered in the second term by $\Delta\varepsilon$), the free energy of the nematic liquid subjected to an electric field has a minimum for a definite orientation of the director relative to the field direction. In nematic liquid crystals with positive dielectric anisotropy ($\Delta\varepsilon > 0$) the stable state exhibiting a minimum of the free energy is characterized by a director alignment along the field. For nematic liquids with negative dielectric anisotropy ($\Delta\varepsilon < 0$) the stable state corresponds to a director orientation perpendicular to the field direction.

If, in the original state, the direction of the director with respect to the field direction does not satisfy the condition of minimum free energy, then a sufficiently strong electric field gives rise to a torque on the liquid crystal which leads to a reorientation of the director. The deformed state possesses a lower free energy than the initial state; in each deformed state there is an equilibrium between the electric

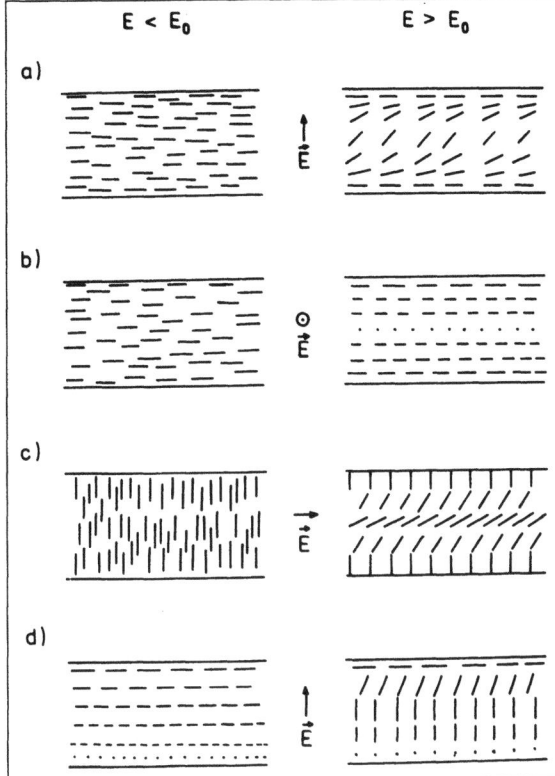

Fig. 2.24. Basic geometries of the dielectric reorientation of nematic liquid crystals ($\Delta\varepsilon > 0$) which are distinguished by different starting orientations with respect to the direction of the applied electric field E (lefthand side). On the righthand side the deformed states above the threshold are displayed.

torque and the restoring elastic torque of the liquid crystal. This equilibrium state is always given by that director configuration through the sample which minimizes the free energy.

Taking into consideration the electric energy $-\frac{1}{2}\varepsilon_0 \Delta\varepsilon (\vec{E}\,\vec{n})^2$ in Eq. 2.29 the distribution of the director through the sample can be calculated.

From Eq. 2.29 expressions for the critical field strength can be derived for which the destabilizing electric torque overcomes the stabilizing restoring elastic torque. There are several geometries in which the field-induced deformations of a nematic liquid crystal can be studied. In Fig. 2.24a–c three basic cases are shown and are characterized by different initial orientation of the director. In all cases the applied electric field is perpendicular to the director. Furthermore, the dielectric anisotropy is positive so that the director tends to align along the field direction.

A) Figure 2.24a: A pure splay deformation results only for a small director displacement. For higher fields the deformation is largely a splay deformation with an admixture of a bend deformation. The threshold field E_0 is given by

$$E_0 = \frac{\pi}{d}\left(\frac{k_{11}}{\varepsilon_o \Delta\varepsilon}\right)^{1/2}, \qquad 2.30$$

where d is the sample thickness.

B) Figure 2.24b: The deformation is a pure twist deformation which starts at a threshold field

$$E_o = \frac{\pi}{d}\left(\frac{k_{22}}{\varepsilon_0 \Delta\varepsilon}\right)^{1/2}. \qquad 2.31$$

C) Figure 2.24c: Small deformations are pure bend deformations. At higher fields the deformation is largely bent with a little splay deformation.
The critical field is

$$E_o = \frac{\pi}{d}\left(\frac{k_{33}}{\varepsilon_0 \Delta\varepsilon}\right)^{1/2}. \qquad 2.32$$

Equations 2.30–2.32 are derived under the simplifying assumptions that the interaction of the molecules with the surface is strong and that the electric conductivity is neglected.

From $E_0 = U_0/d$ (U_0 = threshold voltage) it follows that the threshold voltage is independent of the sample thickness.

Analogous deformations which start at the same critical field strength occur in nematic phases with negative dielectric anisotropy ($\Delta\varepsilon < 0$) when the electric field in each case is perpendicular to the situation drawn in Fig. 2.24.

Figure 2.24d shows schematically the director alignment in a planar-twisted cell where the twist angle is 90°. The threshold for the field-induced reorientation of a nematic phase with positive dielectric anisotropy is given by

$$E_o = \frac{\pi}{d}\left[\frac{k_{11} + \frac{1}{4}(k_{33} - 2k_{22})}{\varepsilon_0 \Delta\varepsilon}\right]^{1/2}. \qquad 2.33$$

Completely analogous deformations, as illustrated in Fig. 2.24a–d, can be generated by magnetic fields. For the theoretical treatment the coupling of the magnetic field H with the diamagnetically anisotropic sample must be considered by the term $-\mu_o \Delta\chi(\vec{H}\cdot\vec{n})^2/2$ which is added to Eq. 2.29. By the same procedure as for the dielectric reorientation formulas of the threshold field, H_o can be derived which are analogous to those of E_o, only $\Delta\varepsilon\,\varepsilon_0$ is replaced by the anisotropy of diamagnetic susceptibility $\Delta\chi\,\mu_0$.

$$H_o = \frac{\pi}{d}\left(\frac{k_{ii}}{\mu_0 \cdot \Delta\chi}\right)^{1/2}. \qquad 2.34$$

It should be noted that a decrease of the anchoring energy causes a decrease of the threshold field. For a tilted director alignment, theoretically, there is no threshold field, the deformation starts at an infinite small field strength. In the case of the dielectric reorientation the anisotropy of electric conductivity and also an inhomogeneous electric field at high field strength ($\Delta\varepsilon > \varepsilon_\perp$) can change the distribution of the director through the bulk of the sample, but the threshold is not affected in this way.

The dynamic of the field-induced deformations can be theoretically treated by the equation of motion of the director which expresses the balance of the elastic and viscous forces and the external electric field. In the special case of a pure twist deformation which is not accompanied by a change in position of the centers of gravity of molecules (in contrast to splay and bend deformations) the equation of motion can be written as:

$$k_{22} \frac{\partial^2 \theta}{\partial z^2} + \varepsilon_0 \Delta\varepsilon \cdot E \sin\theta \cos\theta = \gamma_1 \frac{\partial \theta}{\partial t}, \qquad 2.35$$

where θ is the angle of the local director with respect to the substrates, z is the coordinate perpendicular to the substrate, t is the time, and γ_1 the rotational viscosity. In general, Eq. 2.35 must be solved numerically. For some simplifying assumptions the following equations for the switching times of the dielectric reorientation can be derived by

$$t_{\text{rise}} = \frac{\gamma_1 \cdot d^2}{\varepsilon_0 \Delta\varepsilon} (U^2 - U_0^2)^{-1}, \qquad 2.36$$

where U = voltage and U_0 = threshold voltage, and

$$t_{\text{decay}} = \frac{\gamma_1 \cdot d^2}{\pi^2 \cdot k_{22}}. \qquad 2.37$$

For small deviations from the initial orientation of the director these equations are also valid for bend and splay deformations. Depending on the initial orientation of the director, different elastic constants appear in Eq. 2.37 instead k_{22}, for a planar layer k_{11}, for a homeotropic layer k_{33}, and for a planar twisted layer k_{11} $(k_{33} - 2 k_{22})/4$. It follows from Eqs. 2.36 and 2.37 that the switching times are mainly determined by the rotational viscosity γ_1 and the sample thickness d; for the rise time also $\Delta\varepsilon$ and the driving voltage play important roles.

From the experimentally determined threshold field the corresponding elastic constants are calculated with Eqs. 2.30–2.32, 2.34, provided that the dielectric (or diamagnetic) anisotropy is known. The main problem in the experimental methods for determination of the elastic constants are achieving reproducible anchoring conditions (different surface treatments yield different thresholds) and avoiding a director tilt at the surfaces.

2.6.4 Deformation of Cholesteric Liquid Crystals by Electric Fields

In cholesteric liquid crystals, besides electrohydrodynamic deformations which are connected with the electrical conductivity, also dielectric reorientations can occur. For cholesterics with $\Delta\varepsilon > 0$ the helix axis tends to align perpendicular to the field direction, whereas for cholesterics with $\Delta\varepsilon < 0$ the helix axis parallel to the field direction is energetically favorable. A particular interesting case is the field-induced transition of the cholesteric into nematic state which can be realized for cholesterics with $\Delta\varepsilon > 0$ when the electric field is applied perpendicular to the helix axis. With increasing field strength the pitch of the helix gradually increases (see Fig. 2.25). The critical field strength for the unwinding of the helix is given by

$$E_0 = \frac{\pi^2}{p}\left(\frac{k_{22}}{\varepsilon_0 \Delta\varepsilon}\right)^{1/2}. \qquad 2.38$$

where p is the pitch of the helix.

Using the equation of motion the following approximations for the switching times can be derived:

$$t_{\text{rise}} = \frac{\gamma_1 \cdot p^2}{\pi^4 \cdot k_{22}}\left(\frac{E^2}{E_0^2} - 1\right)^{-1} \qquad 2.39$$

$$t_{\text{decay}} = \gamma_1 \cdot p^2 / k_{22} \cdot \pi^4. \qquad 2.40$$

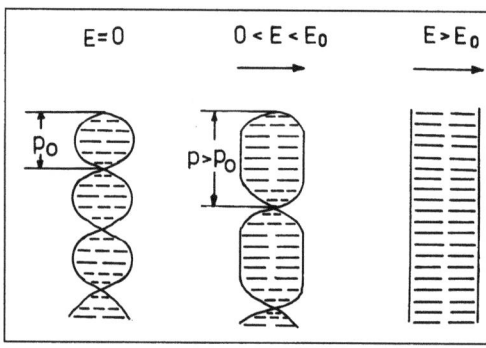

Fig. 2.25. Unwinding of the helix of a cholesteric liquid crystal by an electric field E perpendicular to the helix axis h (schematic representation). E_0 is the threshold field for the transition cholesteric → nematic; p_0 is the pitch of the undistorted state. The dielectric anisotropy of the untwisted nematic state must be positive.

2.6.5 Deformation of Smectic Liquid Crystals (S_A, S_C) by an Electric Field

The smectic layers of a S_A phase can bend easily, which leads to splay deformations of the director. The splay elastic constant k_{11} should be of the same order of magnitude

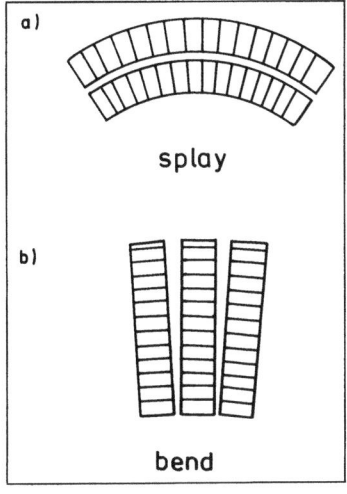

Fig. 2.26. Schematic representation of: a) splay and b) bend deformations in a S_A phase.

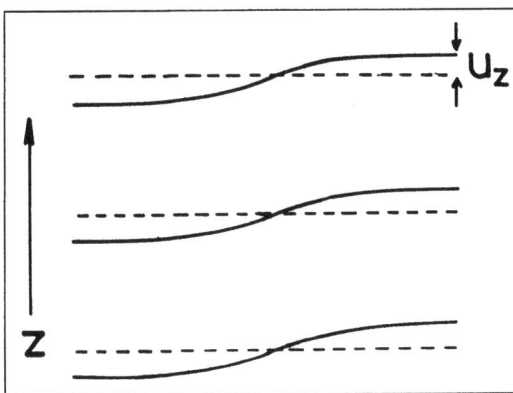

Fig. 2.27. Distortion of a smectic layer (solid line) from its equilibrium state (dashed line).

as in the nematic phase. Otherwise, bend and twist deformations practically cannot occur because, in this case, the thickness of the smectic layers would be changed which requires a high energy (see the schematic representation in Fig. 2.26). From these considerations it can be expected that k_{33} and k_{22} become very high.

The compression of the smectic layers can be described by an elastic constant $\bar{B}$ which is analogous to the compressibility of an isotropic liquid, the dimension of which is energy per volume; the order of magnitude is about $10^{-10} \ldots 10^{-11} \, \text{N} \, \text{cm}^{-2}$. Considering both kinds of deformations the free energy per volume is given by

$$F = \frac{1}{2}\left[k_{11} (\text{div } \vec{n})^2 + \bar{B}\left(\frac{\partial u_z}{\partial z}\right)^2 \right]. \qquad 2.41$$

The z-direction is parallel to the layer normal, u_z is the local displacement of the smectic layers in z-direction (see Fig. 2.27).

As shown by the theory a dielectric reorientation of a planar oriented S_A phase ($\Delta\varepsilon > 0$), in principle, is possible, but contrary to the nematic phase above the threshold the amplitudes of the deformations are expected to be very small because of the very low compressibility of the smectic layers.

Now, some remarks on the S_C phase: The elastic theory of the S_C phase leads to a complicated equation for the free energy density which contains (in first approximation) 10 elastic constants. The distortion of the smectic layers is described by four elastic constants. From these, one constant corresponds to $\vec{B}$ of the S_A phase (dimension energy per volume). Because of the monoclinic symmetry of the S_C phase three elastic constants are necessary to describe the curvature of the smectic layers. Furthermore, four elastic constants are related to the rotation of the director around the layer normal. There are two elastic constants which take into account the coupling between the different types of deformation.

Similar to S_A phases, all deformations which change the distance of the smectic layers require a very high energy. For this reason, in the expression of the free energy density in first approximation only the terms related to the rotation of the director around the layer normal are considered:

$$F = \tfrac{1}{2} B_1 (\text{div } \vec{u})^2 + \tfrac{1}{2} B_2 (\vec{u} \text{ rot } \vec{u})^2$$
$$+ \tfrac{1}{2} B_3 (\vec{k} \text{ rot } \vec{u})^2 - B_{13} (\vec{u} \text{ rot } \vec{u})(\vec{k} \text{ rot } \vec{u}), \qquad 2.42$$

where B_1, B_2, B_3, B_{13} are the elastic constants, and $\vec{u}$ = unit vector parallel to the projection of the director on the smectic plane, and $\vec{k}$ = layer normal (Fig. 2.28).

Deformations based on the rotation of the director around the layer normal are a feature of S_C phases. Field-induced deformations of this type can be experimentally realized by planar aligned S_C phases exhibiting positive dielectric anisotropy. But unlike nematic phases, the planar alignment is not only characterized by a uniform orientation of the director parallel to the substrate plane, but additionally by the orientation of the smectic layer planes (or the layer normals). The orientation of the layer normals can be described by an angle δ between the layer normal and the substrate plane. There are two borderline cases: If $\delta = 0$ the smectic layers are perpendicular to the substrates (book-shelf geometry, see Fig. 2.29a). For $\delta = \theta$ (θ: tilt angle of the S_C phase) the smectic layers are tilted by $90° - \theta$ with respect to the substrate planes (see Fig. 2.29). The general case can be described by $0 < \delta < \theta$.

For the first borderline case ($\delta = 0$) a continuous field-induced dielectric reorientation takes place which is quite similar to that in nematics. The threshold field can be calculated in the same way as known for nematic phases:

$$E_0 = \frac{\pi}{d} \sin\theta \sqrt{\frac{B}{\varepsilon_0 \Delta\varepsilon}}, \qquad 2.43$$

where θ = tilt angle, and B = elastic constant of a one-constant approximation. For $\delta \neq 0$ the threshold field must be calculated numerically; only for special cases are

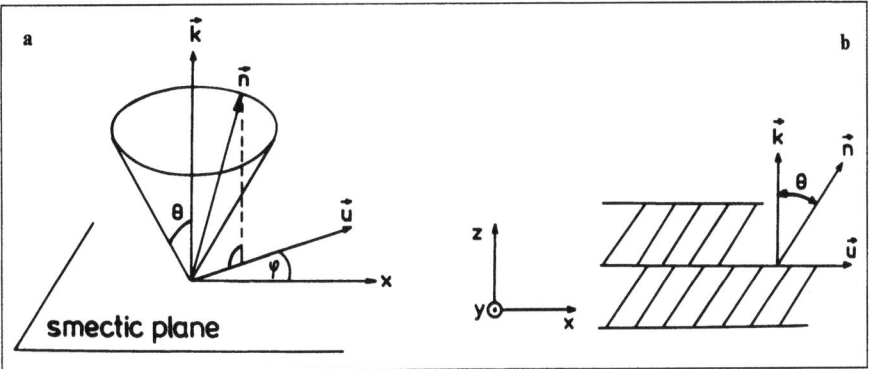

Fig. 2.28. a) Structure model of the S_C phase; $\vec{n}$: director; $\vec{k}$: layer normal; Θ: tilt angle. b) Rotation of the director $\vec{n}$ around the layer normal $\vec{k}$; $\vec{u}$: unit vector oriented parallel to the projection of $\vec{n}$ on the layer plane; φ: rotation angle of $\vec{n}$.

analytical expressions also available. However, unlike the case $\delta = 0$, the field-induced transition is discontinuous, i.e., above the threshold the molecules are displaced by a finite amount with respect to their initial positions. Generally, the threshold is two to four times higher than in the nematic phase of the same substance.

For the switching times in first approximation the following expressions are obtained:

$$t_{rise} = \frac{d^2 \gamma_1}{\varepsilon_0 \Delta\varepsilon \sin\theta} (U^2 - U_0^2)^{-1} \qquad 2.44$$

$$t_{decay} = \frac{\gamma_1 d^2}{\pi \cdot B}, \qquad 2.45$$

where γ_1 is rotational viscosity.

Equation 2.44 is only valid for $\delta = 0$, whereas Eq. 2.45 is not restricted to a special angle δ. It was found that the rise times in S_C and nematic phases are of the same order of magnitude. Otherwise, the decay times in the S_C phase were found to be clearly lower than in the nematic phase of the same substance. This result must be interpreted by a lower rotational viscosity in the S_C phase.

In principle, dielectric reorientations based on the rotation of the director around the layer normal are also possible in the higher ordered tilted smectic phases such as S_I. But the mechanism of the reorientation accompanied by domain wall motions or hysteresis effects is much more complicated, which is obviously due to the pseudohexagonal structure of this phase.

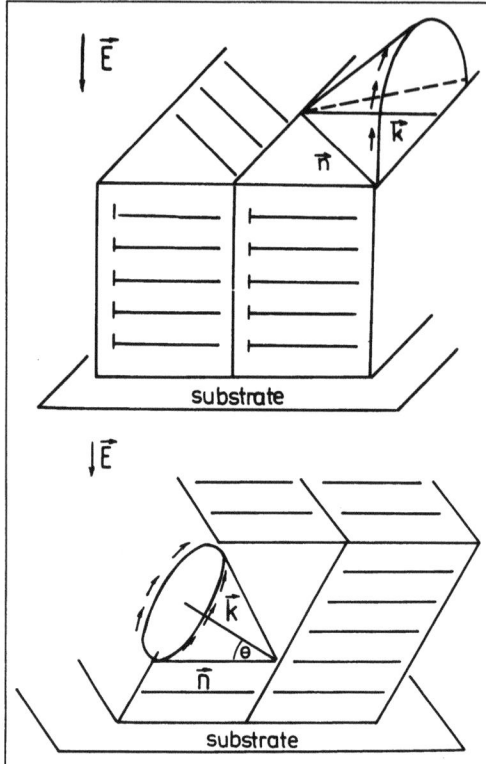

Fig. 2.29. The borderline cases of planar aligned S_C phases a) $\delta = 0$: smectic layers are perpendicular to the substrates; b) $\delta = \Theta$: smectic layers are tilted by $\frac{\pi}{2} - \Theta$ with respect to the substrate plane; δ: angle between layer normal $\vec{k}$ and the substrate plane; $\vec{n}$: director; Θ: tilt angle; $\vec{E}$: direction of the applied electric field.

2.7 Ferroelectric Properties

In S_C phases composed of chiral molecules the symmetry is reduced from C_{2h} to C_2 whereby only a simple two-fold axis remains (S_C^* phase).

If the chiral molecules possess a dipole moment perpendicular to their long axes and if the free rotation around the molecular long axes is hindered, a spontaneous polarization in each smectic layer can result which is directed along the two-fold symmetry axis, i.e., parallel to the layer planes and perpendicular to the tilt plane. The spontaneous alignment of permanent dipoles gives rise to the ferroelectricity within each smectic layer of the S_C^* phase

The chirality of the molecules also leads to a helical structure of the S_C phase, the pitch p of which corresponds to a rotation of the director through an angle of 2π. For this reason, also the direction of the spontaneous polarization rotates from layer to layer, so that for $d \gg p$ (d = sample thickness) the overall polarization is cancelled and in macroscopic S_C^* phases the polarization is absent (Section 1.4.1.1; Fig. 1.26).

The polarization can only be observed if the axial symmetry of the S_C^* phase is disturbed which can be performed in different ways:

A) When the S_C^* phase (the helix axis of which is oriented normal to the substrates) is subjected to a shear along the smectic layers the helix will be distorted and a finite polarization will appear at a right angle to the shear.
B) The distortion and total unwinding of the helix can be achieved by an electric field which is applied perpendicular to the helix. The unwinding of the helix means that the uniaxial S_C^* phase is transformed into the biaxial S_C phase which can be easily detected by conoscopic investigations. Contrary to the field-induced unwinding of the cholesteric phase in the case of the S_C^* phase the electric field couples preferably with the spontaneous polarization. For this reason, in the equation of the free energy (which is the basis of the theoretical description by means of the continuum theory) the electric term $\vec{P}_S \cdot \vec{E}$ must be considered, where P_S is the spontaneous polarization. The critical field for the untwisting of the helix is given by

$$E_0 = \frac{\pi^4 \cdot k_{22}}{4 \cdot p^2 P_S}, \qquad 2.46$$

where p = pitch and k_{22} = twist elastic constant.

C) A third way to suppress the helix is the elastic interaction of the S_C^* phase with the surrounding surface which is realized when the cell thickness is lower or about equal to the pitch (surface-stabilized ferroelectric liquid crystal, SSFLC). By choosing the smectic layers perpendicular to the glass plates (book-shelf geometry) two equally stable director configurations can exist which possess an inverse direction of the spontaneous polarization (see Fig. 2.30). Therefore, by change of the sign of an applied electric field a switching between these states is possible.

Depending on the molecular structure (magnitude and direction of the dipole moment, position and polarity of the chiral center) the spontaneous polarization P_S is found to be between 3 and 1200 nC cm^{-2}.

The switching time, which can be approximately expressed by

$$\tau = \frac{\gamma}{P_S} \cdot E, \qquad 2.47$$

(where γ = rotational viscosity) is about 200 times shorter than that of a dielectric reorientation of nematics (often below a µs). The bistability behavior and the short switching of this electrooptical effect opens new perspectives for practical applications.

It should also be noted that higher ordered tilted smectic phases (S_F, S_I...) composed of chiral molecules can exhibit a helical structure and ferroelectric properties.

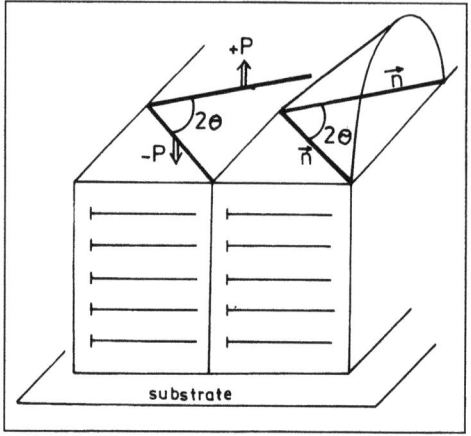

Fig. 2.30. "Book-shelf" geometry of smectic layers of a S_C^* phase. There are two equally stable director orientations $\vec{n}$ which exhibit an inverse direction of the spontaneous polarization P_S.

2.8 Flexoelectric Polarization

In an undistorted nematic liquid crystal the distribution of the molecular dipoles has a non-polar cylindrical symmetry. But if the constituting polar molecules possess a certain assymmetrical shape a macroscopic polarization can occur which is associated with special deformations.

For conical molecules a splay deformation will cause a preferred alignment of the thicker ends in the direction of divergence. If the molecular dipole has a non-vanishing component parallel to the long molecular axis a polarization of the sample arises (see Fig. 2.31a).

For nematic liquids constituting of banana-like molecules with a dipole moment perpendicular to the long molecular axes a bend deformation leads to a polarization of the sample (Fig. 2.31b).

This effect is not identical with the piezoelectric effect of solid crystals where a charge arises on the surface by compression or dilatation. Because in the case of the nematic liquid crystal the physical origin is not translational deformations, as in solids, but rather curvature strains (orientation deformation), the term "flexoelectricity" is used.

The inverse effect of the flexoelectric polarization can be realized when the nematic liquid crystal is subjected to an external electric field. In this way a splay or bend deformation, respectively, is induced by aligning of the permanent dipole moments.

The flexoelectric polarization can be expressed by

$$P = e_{11} \vec{n}(\operatorname{div} \vec{n}) + e_{33}(\operatorname{rot} \vec{n}) \times \vec{n}, \qquad 2.48$$

where e_{11} and e_{33} are the flexoelectric coefficients with the dimensions charge per length. If an electric field is applied on a nematic sample in the expression of the free energy the flexoelectric term $\vec{P} \cdot \vec{E}$ must be considered.

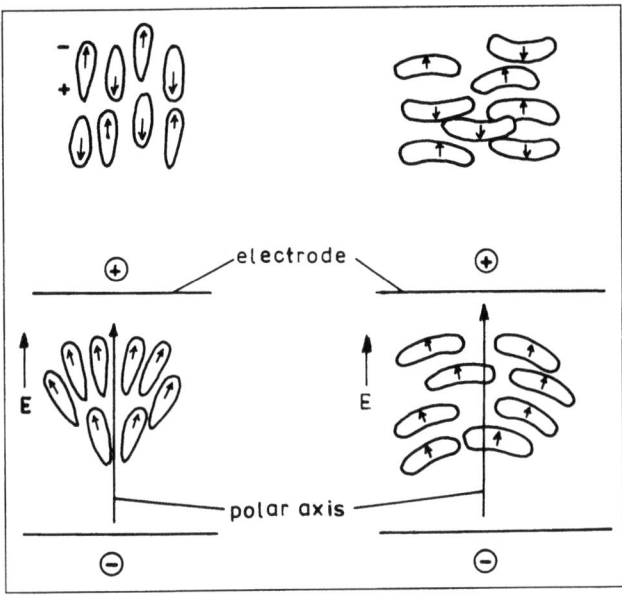

Fig. 2.31. a) Splay deformation of a nematic liquid crystal composed of wedge-shaped molecules which exhibit a longitudinal dipole moment.
b) Field-induced bend deformation of a nematic liquid crystal composed of banana-shaped molecules which exhibit a transverse dipole moment.

2.9 Diffusion Coefficients

In nematic liquid crystals the mobility of the molecules is liquid-like in all directions. Therefore, the diffusion coefficient of the molecules in a nematic phase (self-diffusion), including those of dopand molecules are in the same order of magnitude as in the isotropic liquid. Usual values lie between 0.5×10^{-6} and 6×10^{-6} cm^2 s^{-1}. The structural anisotropy of the nematic phase, however, gives rise to an anisotropy of the diffusion coefficients. It is found that the diffusion coefficient parallel to the director ($D_\parallel$) is larger than that perpendicular to the director ($D_\perp$) (see Fig. 2.32); commonly, the ratio $D_\parallel/D_\perp$ varies between 1.3 and 4. The diffusion is thermally activated with an energy of the order 15...50 kJ mol^{-1} which is comparable with the isotropic liquid state.

In S_A phases the diffusion within the smectic layers is found to be liquid-like and similar to that of nematics, whereas the translational diffusion between the smectic layers largely proceeds by jump processes. The diffusion coefficients are somewhat lower than in the nematic phases; their values are found to be between 1×10^{-7} and 4×10^{-6} cm^2 s^{-1}. The temperature dependence of $D_\parallel$ and $D_\perp$ follows the Arrhenius law, as in nematics.

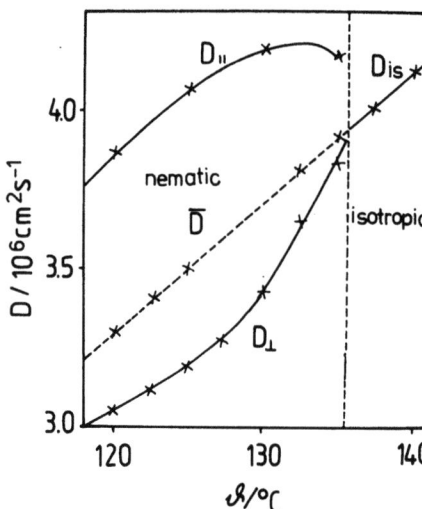

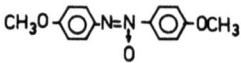

Fig. 2.32. Self-diffusion coefficients in the nematic and isotropic phase of 4,4'-di-methoxyazoxy benzene. $D_\parallel, D_\perp$: diffusion coefficients parallel and perpendicular to the director, respectively; D_{is}: diffusion coefficient of the isotropic liquid; $\bar{D}$: average diffusion coefficient of a non-oriented nematic sample [27].

Because the activation energy of $D_\parallel$ is higher than for $D_\perp$ with decreasing temperature the ratio $D_\parallel/D_\perp$ can become smaller than unity.

In the higher ordered smectic phases only a few measurements of the diffusion coefficient have been performed. In S_B phases similar to S_A phases $D_\parallel/D_\perp < 1$ is found. The diffusion coefficient $D_\parallel$ which corresponds to the translational diffusion between the smectic layers is in the order of 3×10^{-8} cm^{2}s^{-1}.

2.10 Electric Conductivity

Like organic liquids, the liquid crystals exhibit an intrinsic electric conductivity which is mainly due to ionic impurities. The conductivity of carefully purified liquid crystals commonly lies between 10^{-12} and 10^{-14} Ω^{-1} cm^{-1}. In non-purified samples the conductivity can be higher by several decades because of the higher concentration of ionic impurities.

Approximately, the equivalent electrolytic conductivity Λ as function of the concentration c can be expressed by

$$\Lambda = (K/c)^{1/2}[\Lambda_\infty - A(K \cdot c)^{1/4}], \qquad 2.49$$

where Λ_∞ = equivalent conductivity at infinite dilution, K = dissociation constant, and A = constant of the Onsager theory.

For weak electrolytes such as liquid crystals the degree of dissociation and also the dissociation constant K is very low, so that usually the second term in Eq. 2.49 can be neglected. From this it follows that the equivalent electrolytic conductivity Λ is proportional to $1/\sqrt{c}$ whereas the specific conductivity $\sigma = c \cdot \Lambda/1000$ is proportional to $\sqrt{c}$.

The electric conductivity σ of an electrolytic solution depends on the ion concentrations of the ions c_i, on their charge number z, and on their mobility u_i:

$$\sigma = \sum z_i \cdot c_i \cdot F u_i, \qquad 2.50$$

where F is the Faraday constant.

Otherwise, according to Einstein's law, the ionic mobility is proportional to the diffusion coefficient:

$$u \sim D. \qquad 2.51$$

Because the diffusion in liquid crystals is anisotropic, an anisotropy of electric conductivity results which can be expressed by

$$\Delta\sigma = \sigma_\parallel - \sigma_\perp, \qquad 2.52$$

where $\sigma_\parallel, \sigma_\perp$ is electric conductivity parallel and perpendicular to the director, respectively.

Using Eq. 2.51, the ratio $\sigma_\parallel/\sigma_\perp$ can be written as

$$\sigma_\parallel/\sigma_\perp = \frac{D_\parallel}{D_\perp}. \qquad 2.53$$

In general, anisotropy of the conductivity of nematic liquid crystals is positive ($\sigma_\parallel - \sigma_\perp > 0$; $\sigma_\parallel/\sigma_\perp > 1$). With decreasing temperature $\sigma_\parallel - \sigma_\perp$ ($\sigma_\parallel/\sigma_\perp$) increases corresponding to the increasing orientational order. As expected, the ratio $\sigma_\parallel/\sigma_\perp$ strongly depends on the size and shape of the ionic charge carriers dissolved in the nematic solvent. This is obvious from Fig. 2.33, where the temperature dependence of $\sigma_\parallel/\sigma_\perp$ for various electrolytes dissolved in the same nematic solvent is shown. It is not surprising that small spherical ions (curve 1, 2) possess a relatively low conductivity anisotropy, whereas dimethyl-ethyl-octadecylammonium-4-hexyloxybenzoate (curve 4) exhibit a relatively high $\sigma_\parallel/\sigma_\perp$ because of the rod-like shape of the ions.

In smectic phases, i.e., S_A or S_B, the conductivity anisotropy is negative ($\sigma_\parallel - \sigma_\perp < 0$) because the diffusion of the ionic impurities perpendicular to the director is easier than the translational diffusion parallel to the director (perpendicular to the smectic layers). A typical example is shown in Fig. 2.34. It is seen that $\sigma_\perp$ increases at the transition $N \to S_A$ and $S_A \to S_B$, whereas $\sigma_\parallel$ monotonically decreases with decreasing temperature. It is remarkable that, even in the nematic phase, the conductivity anisotropy changes the sign on approaching the nematic $\to S_A$ transition. This is obviously due to pretransitional effects and the successive formation of smectic clusters (cybotactic groups).

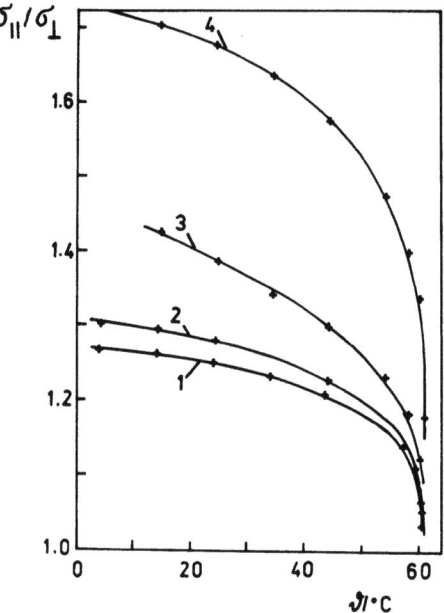

Fig. 2.33. The ratio $\sigma_\parallel/\sigma_\perp$ in dependence on the reduced temperature T/T_{NI} (T_{NI} = clearing temperature) for different electrolytes dissolved in the nematic phase of a commercial mixture (Merck ZLI 599).
1: tetraethylammoniumperchlorate;
2: tetrabutylammoniumperchlorate;
3: trimethyl hexadecylammoniumbromide;
4: dimethyl ethyl octadecylammonium-p-hexyloxybenzoate [7].

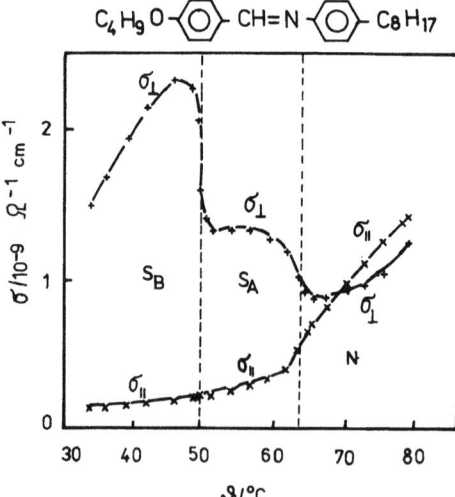

Fig. 2.34. The temperature dependence of the electrical conductivity parallel ($\sigma_\parallel$) and perpendicular to the director ($\sigma_\perp$) in the nematic, S_A and S_B phase of 4-n-butyloxy-benzylidene 4-n-octylaniline [15].

2.11 Electrohydrodynamic Instabilities

Whereas for the study of dielectric reorientations in liquid crystals the conductivity should be as low as possible, in the case of electrohydrodynamic deformations the conductivity and its anisotropy play an important role because an electrical current coupled with a mechanical flow is an essential condition for this type of instability.

The most studied electrohydrodynamic instabilities in liquid crystals are the Williams–Kapustin domains. They appear when a DC or low-frequency AC field is applied to a planar oriented nematic sample (director uniformly parallel to the substrate planes) having negative dielectric anisotropy and a sufficient electrical conductivity. Above a critical field strength a stationary stripe pattern arises whereby the stripes are perpendicular to the original director alignment (see Fig. 2.35). These stripes are only visible for light polarized parallel to the direction of the director. Furthermore, the period of the stripes is approximately equal to the sample thickness and independent of the applied voltage. The effect is associated with fluid flow which can be easily seen from the motion of suspended dust particles. With increasing voltage the domain pattern becomes more complicated and is successively destroyed.

At higher voltages the medium goes over to a turbulent state which strongly scatters the incident light. This field-induced light scattering (called dynamic scattering mode) can, in principle, be applied for display devices.

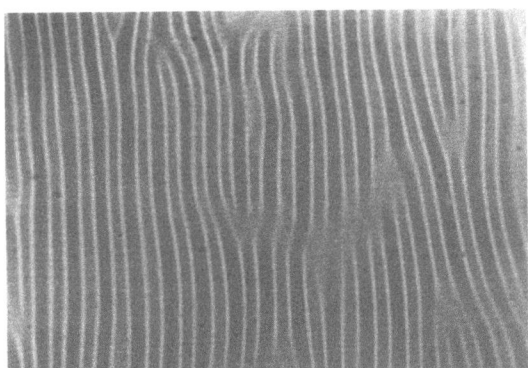

Fig. 2.35. Williams–Kapustin domains for a nematic liquid crystal with negative dielectric anisotropy just above the threshold voltage. The stripes are perpendicular to the original orientation of the director.

The mechanism of the formation of the Williams–Kapustin domains can be explained in the following way:

For small fluctuations of the local director the anisotropy of the conductivity (the mobility of the ions along the director is greater than that perpendicular to it) produces a charge separation which is schematically shown in Fig. 2.36. The applied electric field exerts forces on the charged volume, giving rise to a flow. This flow in alternating directions leads to a viscous torque on the local directors in such a direction that the deformation increases.

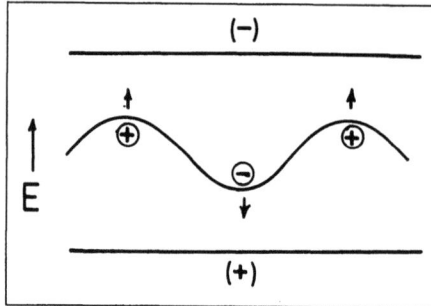

Fig. 2.36. Occurrence of space charges in an applied electric field caused by bend fluctuations. The bend of the director is designated as a solid line.

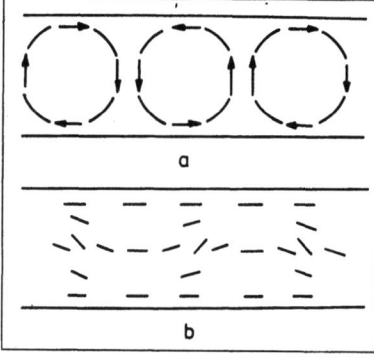

Fig. 2.37. a) Flow pattern associated with the Williams–Kapustin domains; b) Local orientation of the director giving rise to the domain pattern [after 19].

The deformation is reinforced by the dielectric torque due to the transverse field created by the space-charge distribution. On the other hand, there are two counteracting effects which try to restore the uniform planar alignment; firstly the elastic torque which is due to the surface interaction, and, secondly, the dielectric torque because of the negative dielectric anisotropy. Above the threshold the deforming torques overcome the stabilizing torques, giving rise to the periodic deformation of the nematic layer. As illustrated in Fig. 2.37 the periodic deformation is caused by a liquid flow in cylindrical motion perpendicular to the domains' stripes which correspond to the vortices of the motion. Because the direction of the director is identical to the direction of the optical axis, the periodic change of the local optical axes leads to a periodic change of the refractive index so that the nematic liquid acts as a lattice of cylindrical lenses. For light polarized perpendicular to the original direction of the director there is no change of the refractive index, therefore, in this case the domain pattern cannot be observed.

From the balance between the stabilizing and destabilizing torques the threshold of the electrohydrodynamic instabilities can be derived:

$$U_0 = \frac{2\pi d}{\lambda} \sqrt{\frac{k_{33}}{\varepsilon_0 \varepsilon_\| \frac{\Delta \varepsilon \sigma_\perp}{\varepsilon_\| \sigma_\|} - \eta' \frac{\varepsilon_\perp}{\varepsilon_\|} - \frac{\sigma_\perp}{\sigma_\|}}}, \qquad 2.54$$

where d = sample thickness, k_{33} = bend elastic constant, λ = period of the pattern, and η' = effective viscosity.

Because the period λ of the Williams–Kapustin domains is nearly equal to the sample thickness d the threshold voltage U_0 is approximately independent of the sample thickness.

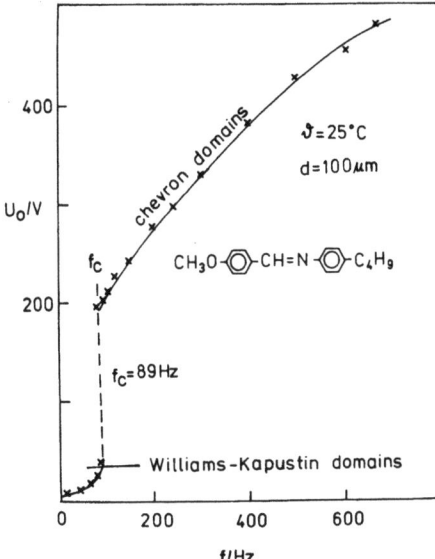

Fig. 2.38. Frequency dependence of the threshold voltage for the occurrence of electrohydrodynamic instabilities in the nematic phase of 4-methoxybenzylidene 4-n-butylaniline ($\vartheta = 25\,°\mathrm{C}$, $d = 100\,\mu\mathrm{m}$, $f_c = 89\,\mathrm{Hz}$). I: Williams–Kapustin domains; II: Chevron domains [16].

The threshold voltage U_0 of the Williams–Kapustin domains generally increases with increasing frequency. As a critical frequency f_c is approached a divergence of the threshold voltage is observed (see Fig. 2.38). According to the theory, the frequency dependence of U_0 can be described by

$$U_0^2(f) = U_0^2[1 + (\xi^2 - 1)(f/f_c)^2]/[1 - (f/f_c)^2], \qquad 2.55$$

where f is the frequency and U_0 is the threshold for the DC field. The parameter ξ^2 is given by

$$\xi^2 = \left(1 - \frac{\sigma_\perp \varepsilon_\|}{\sigma_\| \varepsilon_\perp}\right)\left(1 + \frac{\varepsilon_\|}{\Delta \varepsilon} \eta'\right), \qquad 2.56$$

where η' is the viscosity parameter.

The critical frequency f_c can be calculated according to

$$f_c = \frac{\sqrt{\xi^2 - 1}}{2\pi\tau},\qquad 2.57$$

where τ is the dielectric relaxation time:

$$\tau = \varepsilon_0 \varepsilon_\| / \sigma_\|.\qquad 2.58$$

This means that the critical frequency is directly proportional to the conductivity and thus can be varied by doping with ionic impurities.

Above f_c other types of domains appear which are called "chevron domains" because of their typical pattern (see Fig. 2.39). The distance between the striations is smaller than the sample thickness and depends on the frequency. Furthermore, the chevron domains are not stationary, rather, they oscillate. The threshold of the chevron domains is determined by a critical field strength rather than a critical

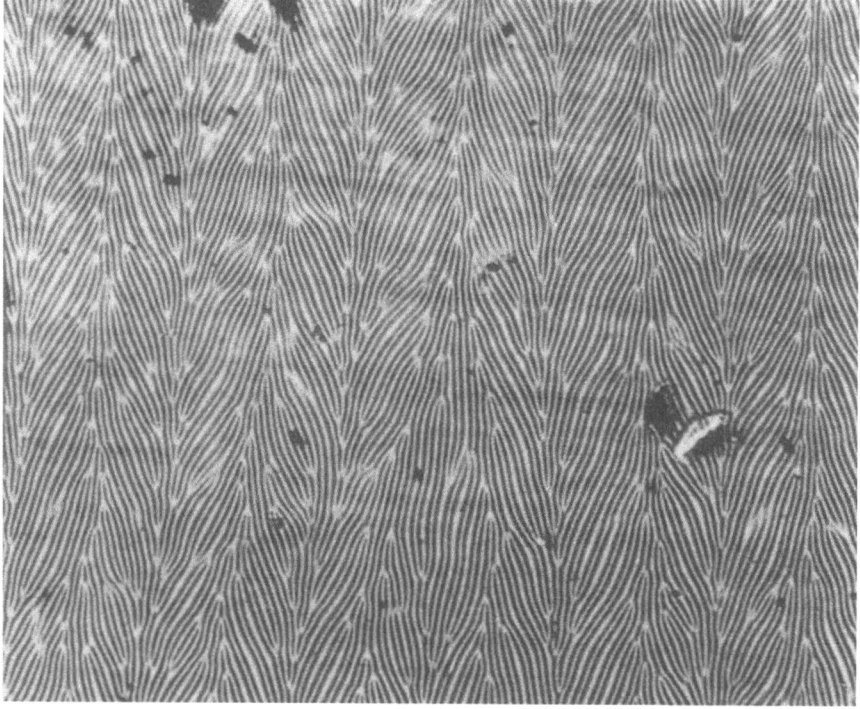

Fig. 2.39. Chevron pattern of oscillating domains in a nematic liquid with negative dielectric anisotropy.

voltage, and it varies with the square root of the frequency. The chevron pattern also becomes turbulent at about twice the threshold field.

As shown earlier, below f_c the distortion of the director is static and the space charge oscillates with the frequency of the applied field. Above f_c – in the regime of the chevron domains – the inverse situation appears whereby the distribution of the space charge is approximately stationary while the local director fluctuates.

Electrohydrodynamic instabilities in nematics can occur also for other signs of the dielectric and conductivity anisotropy depending on the initial orientation of the director. Furthermore, electrohydrodynamic domain patterns of different kind can be observed also in cholesteric phases as well as in S_C phases. It is interesting that, in nematic and cholesteric liquid crystals, hydrodynamic instabilities can arise under thermal gradients which are similar to the Benard instability of isotropic liquids. But, in the nematic or cholesteric phase, the threshold thermal gradient is much lower than in the isotropic phase.

2.12 Viscosity

The flow of an isotropic liquid through a cappillary is described by Poiseuille's law, according to which the rate of flow v is given by

$$v = \Delta p r^2 / 8l\eta, \qquad 2.59$$

where r is the radius and l the length of the capillary; Δp is the constant difference in pressure exerted by the hydrostatic pressure of the liquid, and η is the viscosity. In nematic liquid crystals the flow and, therefore, the effective viscosity depends on the angle between the director and the direction of flow and on the direction of the velocity gradient. There are three limiting cases which are illustrated in Fig. 2.40.

The alignment of the director is controlled by a sufficiently high magnetic field H because the magnetic torque must be much higher than the torque exerted by the shear flow. The viscosity η_2 is measured when the elongated molecules are aligned in the direction of the flow, but are perpendicular to the velocity gradient. The viscosity coefficients η_3 and η_1 correspond to a director alignment perpendicular to the flow direction, but in the case of η_3 the molecular director is perpendicular; in the case of η_1 parallel to the velocity gradient (Fig. 2.40). The measurements of the viscosity coefficients which correspond to the three limiting cases can be performed by an oscillating plate viscosimeter. In this method the one bounding plate is fixed while the other is moved to create a shear (see Fig. 2.40). The moving plate oscillates at very low frequency. From the damping of the oscillations by the viscuos medium the viscosities can be calculated. Figure 2.41 presents the results for the nematic phase of 4-methoxybenzylidene 4-n-butylaniline. It is seen that $\eta_1 > \eta_3 > \eta_2$ and that the η_3-curve is practically the continuation of the corresponding curve in the isotropic liquid state. Otherwise, the viscosity coefficient η_2 possesses a minimum value; at the clearing temperature, η_2 is even lower than the viscosity coefficient of the isotropic liquid. This fact is reasonable because the flow of parallelly oriented, elongated

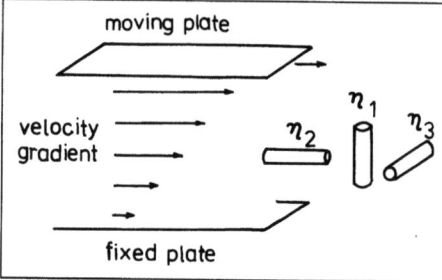

Fig. 2.40. The three principal viscosity coefficients η_1: director perpendicular to the flow and parallel to the velocity gradient; η_2: director parallel to the flow and perpendicular to the velocity gradient; η_3: director perpendicular to the flow and perpendicular to the velocity gradient. Principle of the oscillating plate viscosimeter. The shear is induced by the moving upper plate and the director alignment is established by a magnetic field.

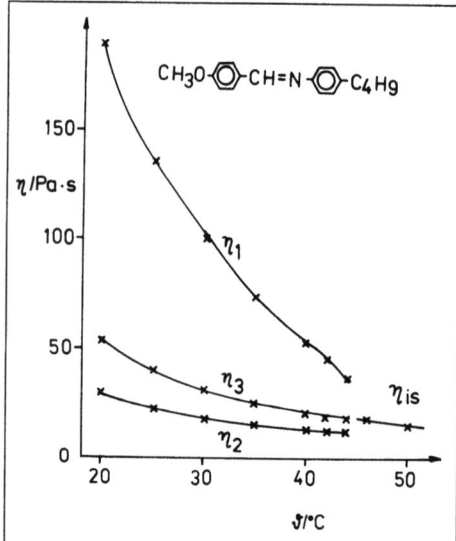

Fig. 2.41. Temperature dependence of the principal viscosity coefficients in the nematic (η_1, η_2, η_3) and isotropic phase (η_{is}) of 4-methoxybenzylidene 4'-n-butylaniline [11].

molecules along the flow direction should be easier than the flow of the randomly oriented molecules of an isotropic liquid. Because the flow generally leads to an alignment of the director in flow direction by the flow of a nematic liquid through a capillary without orienting magnetic field, mainly the minimum viscosity coefficient η_2 is measured. In general, the viscosity increases with increasing number of rings or increasing length of terminal alkyl chains of the liquid-crystalline compounds. Also, lateral branches lead to an enhancement of the viscosity, whereas substituting of aromatic rings by alicyclic rings reduces the viscosity. It is apparent that the magnitude of the viscosity is not only affected by attractive intermolecular forces, but also by steric factors (length and breadth of the molecules, lateral branches).

As described in Section 2.6.3, the rotational viscosity γ_1 plays an important role in the dynamic of the field-induced reorientations. In principle, γ_1 is available from the switching times of the dielectric reorientations according to Eqs. 2.36, 2.37, 2.39, 2.40,

2.44, 2.45. But only a pure twist deformation gives exact values, because in this case there is no hydrodynamic flow and the molecules rotate without any translational motion. The rotational viscosity of technically important nematic liquid crystals is in the range of 0.02 Pa s to 0.5 Pa s.

It is remarkable that γ_2 (and γ_1) diverges as the transition temperature nematic $\to S_A$ is approached, whereas γ_3 does not show anomalies. This behavior indicates the existence of a pretransitional smectic order.

In the smectic state an additional mechanism exists which is characterized by the permeation of the molecules normal to the layers. This effect strongly enhances the viscosity γ_2 of the S_A phase.

On the other hand, it should be emphasized that the rotational viscosity of a S_C phase which is related to the rotation of the molecules around the layer normal is lower than that of a nematic phase of the same substance.

References

1. Blinov LM (1983) Electro-optical and magneto-optical properties of liquid crystals. Wiley & Sons, Chichester, New York, Brisbane, Toronto, Singapore
2. Chandrasekhar S (1977) Liquid Crystals. Cambridge University Press, Cambridge
3. Gasparoux H, Hardouin F, Achard MF (1975) Magnetic properties of smectic mesophases. Pramana Suppl 1 : 215
4. Gennes PG de (1974) The Physics of Liquid Crystals. Clarendon Press, Oxford
5. Hauser A, Selbmann C, Demus D, Grande S, Petrov AG (1983) Order parameter and molecular polarizabilities of liquid crystals with nematic and smectic phases. Mol Cryst Liq Cryst 91 : 97
6. Hauser A, Selbmann C, Rettig R, Demus D (1986) Physical Properties of Liquid Crystalline 5-n-Hexyl-2-(4-n-alkyloxyphenyl)-pyrimidines. Cryst Res Technol 21 : 685
7. Heppke G, Schneider F (1976) Anisotropie der elektrischen Leitfähigkeit verschiedener Elektrolyte in einem nematischen Flüssigkristall. Z Naturforsch 31a : 611
8. Heppke G, Kitzerow H, Oestreicher F, Quentel S, Ranft A (1988) Electrooptics effects in a non-polar nematic discotic liquid crystal. Mol Cryst Liq Cryst Lett 6 : 71
9. Jeu WH de (1978) Liquid Crystals. Solid State Physics Suppl No 14. Academic Press, New York
10. Jeu WH de (1980) Physical Properties of Liquid Crystalline Materials. Gordon & Breach, New York
11. Kneppe H, Schneider F (1981) Determination of the Viscosity Coefficients of the Liquid Crystal MBBA. Mol Cryst Liq Cryst 65 : 23
12. Kresse H, Selbmann C, Demus D, Buka A, Bata L (1981) Reorientation of the Long Molecular Axis in Smectic Phases. Cryst Res Technol 16 : 1439
13. Lockhart IT, Allender DW, Gelerinter E, Johnson DL (1979) Investigations of the indices of refraction. Phys Rev A 20 : 1655
14. Meier G, Saupe A (1966) Dielectric relaxation in nematics. Mol Cryst 1 : 515
15. Mircea-Roussel A, Leger L, Rondolez F, Jeu WH de (1975) Measurements of transport properties in nematics and smectics. J Phys (Paris), 36, Colloq : C1-93
16. Orsay Liquid Crystal Group (1970) Hydrodynamic instabilities in nematics under a.c. electric fields. Phys Rev Lett 25 : 1642

17. Pelzl G, Sackmann H (1971) Birefringence and polymorphism of liquid crystals. Symp Farad Soc 5 : 68
18. Pelzl G, Sackmann H (1973) Die Doppelbrechung der cholesterinischen und smektischen Modifikationen der homologen n-Fettsäurecholesterylester. Z Phys Chem (Leipzig) 254 : 354
19. Penz PA (1971) Order parameter distribution for the electrohydrodynamic mode of nematic. Mol Cryst Liq Cryst 15 : 141
20. Rettig R (1985) Die Messung der optischen Anisotropie an nematischen Flüssigkristallen. Thesis. Halle/Saale
21. Schad H, Baur G, Meier G (1979) Elastic constants and diamagnetic anisotropy of p-disubstituted phenylcyclohexanes (PCH). J Chem Phys 70 : 2770
22. Schadt M (1972) Dielectric Properties of Some Nematic Liquid Crystals with Strong Dielectric Anisotropy. J Chem Phys 56 : 1494
23. Shashidhar R, Chandrasekhar S (1975) Pressure studies on liquid crystals. J Phys (Paris) 36 : 1–49
24. Smalla KH (1983) Thermodynamische und optische Untersuchungen der Phasenumwandlungen flüssiger Kristalle in binären Systemen, Thesis, Halle/Saale
25. Wiegeleben A, Richter L, Deresch J, Demus D (1980) Calorimetric Investigations of Some Homologous Series with a High Degree of Smectic Polymorphism. Mol Cryst Liq Cryst 59 : 329
26. Wiegeleben A, Demus D (1982) Calorimetric Investigations in Liquid Crystalline TBAA. Cryst Res Technol 17 : 161
27. Yun CK, Fredrickson AG (1970) Anisotropic mass diffusion in liquid crystals. Mol Cryst Liq Cryst 12 : 73

3 Liquid Crystalline Polymers

R. Zentel

Introduction

Considering the possibility of liquid crystalline (LC) polymers at first, they seem to be difficult to achieve, because polymer chains usually adopt a statistical coil conformation, whereas LC-phases possess an orientational and/or a positional long-range order. Nevertheless, polymer chains can be combined with liquid crystallinity in two ways (see Fig. 3.1).

1) Very generally every polymer chain is formanisotropic (much longer than thick). If, therefore, the polymer chain is rigid (or more quantitatively: if the persistence length is long enough), then the polymer chain may act as a mesogenic element itself (the so-called LC-main chain polymers) (see Fig. 3.1a). All polymers with stiff chains can therefore be expected to form preferably nematic phases [1] (so-called thermotropic main chain LC-polymers; see Section 3.1.1). Often, these phases can, however, not be observed directly, because the expected melting temperatures of rigid polymers are far above the temperatures of decomposition (above 300 °C). In these cases the addition of a suitable solvent can lower the melting temperature and allow the observation of the LC-phase (so-called lyotropic LC main chain polymers; see Section 3.1.2).

2) Alternatively flexible polymer chains can be functionalized with formanisotropic mesogens known from low molar mass liquid crystals [2, 3]. If a decoupling element, the so-called flexible spacer, is used to link the mesogens as side groups to the polymer chain, then LC-phases can be observed (so-called LC-side chain polymers; see Section 3.2). In these cases the orientation of polymer chains and mesogenic side groups are largely decoupled from each other, and both subsystems are able to follow their inherent orientational tendencies. The properties of LC-main chain or side chain polymers are rather different. LC-main chain polymers were theoretically predicted and are well understood. LC-side chain polymers were first prepared by experimentalists, but their theory is still in the early stages. In main chain polymers the polymer chain is the mesogenic element itself. Therefore, the role of the polymer chain for the occurrence of the LC-phase is evident. Side chain polymers, however, are composed of two more or less independent subsystems (the mesogens and the polymer chains). While the ordering of the mesogens and their phase sequence is very close to that of low molar mass liquid crystals, the exact role of the polymer chain is not yet clear.

The structure and properties of the different types of LC-polymers will be discussed below in the following sequence. After Section 3.1 about LC-main chain polymers, LC-side chain polymers will be described (Section 3.2). Thereafter LC-polymers with

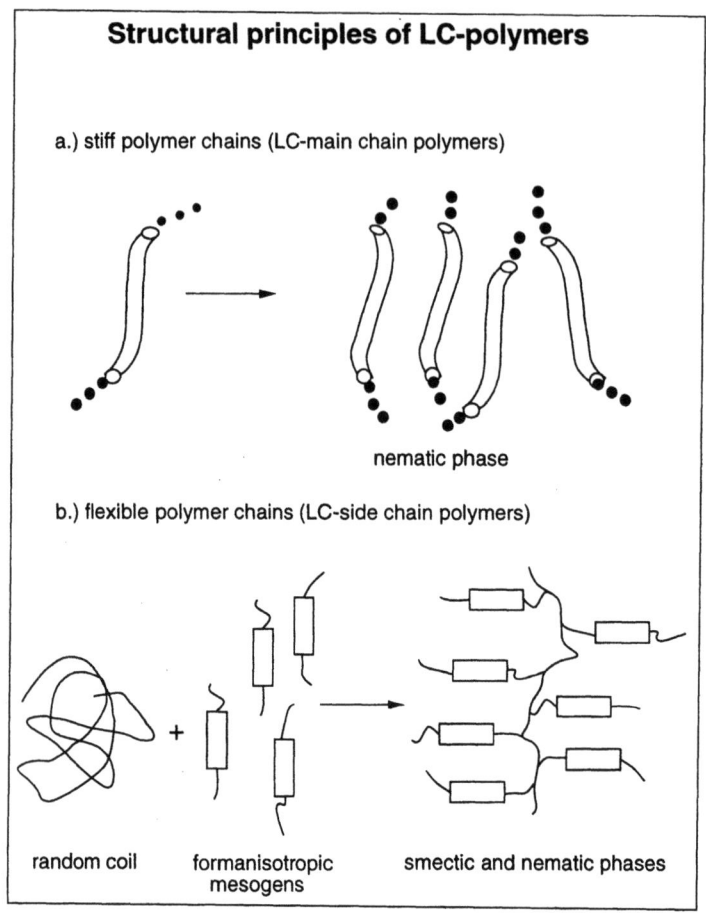

Fig. 3.1. Two alternative routes to LC-polymers start from stiff (a) or flexible polymer chain (b).

discotic phases will be presented independent of their molecular structure (Section 3.3), followed by a section about LC-elastomers (Section 3.4).

3.1 Liquid Crystalline Main Chain Polymers

3.1.1 Thermotropic Main Chain LC-Polymers

As already pointed out in the introduction, polymers with stiff chains are inherently thermotropic, but their melting points are often above the decomposition temperature (T_{dec}) (see also Fig. 3.4). In order to bring the LC-phase to an accessible temperature range, it is therefore necessary to disturb the perfect packing in the

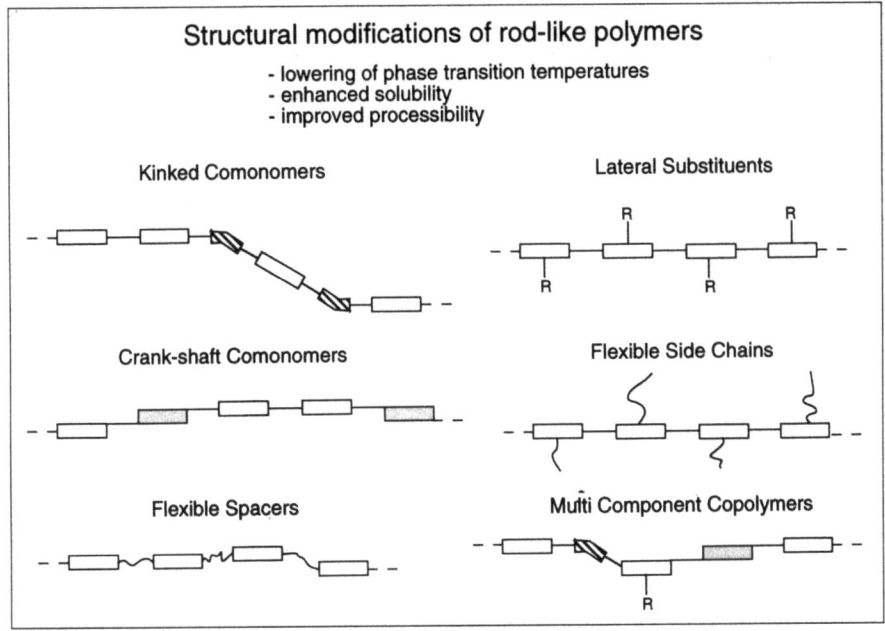

Fig. 3.2. Structural modification of rod-like polymers.

crystalline phase [4–10]. Some possibilities to do this are presented in Fig. 3.2. Fully aromatic polymers of this type are technologically important as thermoplastic polyesters (see Section 3.1.1.1). Their most important properties are: chemical stability, poor solubility, high melting temperatures, and excellent mechanical properties. Polymers with flexible alkylspacers have been, on the other hand, intensively investigated for scientific purposes (see Section 3.1.1.2). They possess their liquid crystalline phases at low temperatures and are well soluble.

3.1.1.1 Thermotropic LC-Polyesters

Some of the structures used to prepare thermotropic aromatic LC-polyesters are presented in Table 3.1. Bulky lateral substituents (polymer 1), crankshaft (polymer 2) or kinked (polymer 3) monomers or short alkyl spacers (polymer 4) are used to lower the melting temperature to about 300 °C. The clearing temperature of all these polymers is still above the decomposition temperature. Their rheological properties are similar to lyotropic LC-polymers (cf. Section 3.1.2). However, they are melt processable. In this respect, it is very important that the viscosity is very low in the nematic phase, much lower than expected for an isotropic polymer melt from comparable polymers. This allows an easy processing of the thermotropic polyesters at high temperatures to prepare bulk samples (plates or more complex three-dimensional structures) and not only fibers as from lyotropic LC-polymers. The

Table 3.1. Examples for aromatic LC-polyesters [5].

Structural units	Mole ratios	$T_m[°C]$
1	1 : 1	340
2	3 : 1	302
3	5 : 3.5 : 1.5	
4	2 : 2 : 6	260

desired material properties are obtained in the ordered crystalline phase. The mechanical properties are excellent in the direction of flow; due to the high melting temperature they can be used at high temperature without a loss of the mechanical properties and their poor solubility makes them inert to all common solvents.

The details of the processing of these polymers into good bulk samples is, however, still unsolved. The orientation of the polymers in bulk samples has to match the direction of maximum strain. Otherwise, the sample will fail, because the mechanical properties perpendicular to the director are rather poor. For more complex structures of the bulk sample, there is still no solution of this problem. One possibility to circumvent this problem is to fill the LC-polyesters with short glass fibers. These short glass fibers prevent an orientation during processing (flow). In addition, they lead to a macroscopic averaging of the anisotropic properties of the LC-matrix, producing a macroscopically isotropic material. The advantage of aromatic LC-polyesters is in this case no longer their liquid crystallinity (in fact, this is more a drawback), but their high melting temperature, their chemical stability, and their insolubility (inertness) in common solvents.

3.1.1.2 Semiflexible Main Chain LC-Polymers

Main chain polymers, which combine flexible spacers with lateral substituents like polymer 6 (see Table 3.2), form thermotropic LC-phases at moderate temperatures. These polymers lack the high tensile strength and the chemical resistivity of fully aromatic polyesters. However, they have been well investigated for scientific purposes, because both the liquid crystalline and the isotropic phase is well accessible.

These polymers are expected to behave like flexible polymers when brought into an isotropic melt or solution. On the other hand, polyester 6 of Table 3.2 does form a nematic mesophase. Hence, this material must undergo major conformational changes when going from the disordered to the liquid crystalline state. The obvious consequence of this is the strong change of entropy at the nematic-to-isotropic transition. This is shown for polyester 6 as function of the number n of the carbon atoms in the spacer in Fig. 3.3 [11]. Also, the order parameter S at $T = 0.98 \cdot T_{ni}$ exhibits a strong alternation with n.

These data demonstrate that the spacer does not merely play the role of a solvent, but takes part actively in the ordering process [12, 13]. Thus, it is now well established that the nature of the spacer groups is of great importance for the stability of the nematic state. If n is an odd number, the subsequent rod-like entities are not collinear if the spacer assumes an all-trans conformation. Therefore, more conformational trans-gauche inversions have to be introduced for an odd number n of methylene units than for the even member of the series in order to align the calamitic units along the nematic director. Therefore, there is a drastic reduction of ΔH_{ni} and ΔS_{ni} when changing the methylene spacer to the more flexible ethyleneoxy spacer. Deuterium NMR-spectroscopic investigations demonstrated that the conformational characteristics of the polymer are very similar to the result found for "twin-dimers," i.e., on model oligomers consisting of two rod-like mesogenic units connected by a single spacer group [14].

Table 3.2. Examples for semiflexible main chain LC-polymers [11].

No.	n	R	phase transitions [°C]
5	10	H	k 216 n 265 i
6	10	CH$_3$	k 118 n 162 i

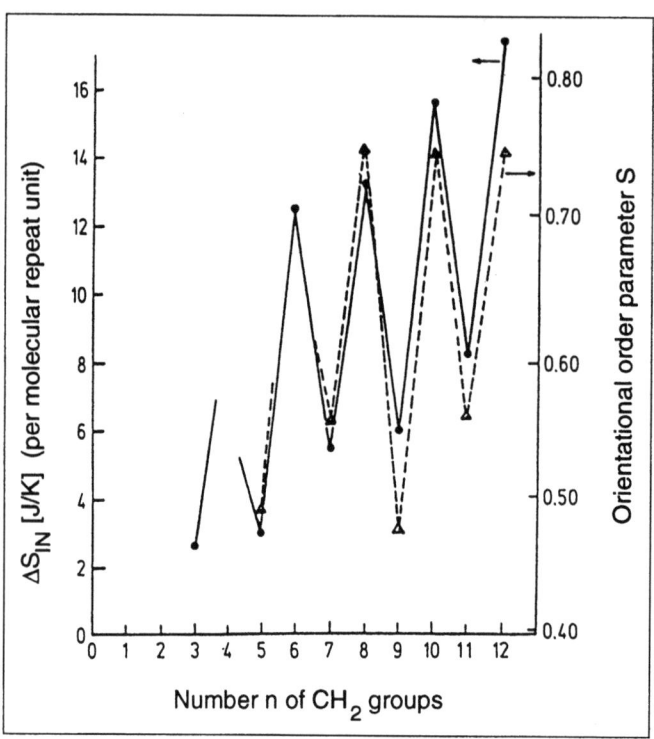

Fig. 3.3. Odd-even effect in semiflexible LC-main chain polymers [11] measured for a homologous series corresponding to polymer 6 (Table 2) with differing length of the alkylspacer (*n*: number of carbon atoms in the alkyl spacer). Left axis: transition entropy nematic/isotropic; right axis: orientational order parameters.

The foregoing discussion has shown that liquid crystalline ordering in polyesters of the type 5–6 requires conformational rearrangements as well. Therefore, it is expected that the overall dimensions of the chain expressed in terms of the radius of gyration will increase when going from the isotropic to the nematic state. A recent study employing small-angle neutron scattering [15] indeed revealed a strong extension of the chains along the preferred direction in the nematic phase. Thus, these findings clearly reveal the strong coupling between orientational and conformational order in LC-polymers with flexible spacers in the main chain.

3.1.2 Lyotropic Main Chain LC-Polymers

As already pointed out, polymers with very stiff chains (rigid rod polymers) are expected to form thermotropic LC-phases as a single component [4–10, 16]. The melting temperature of the crystalline phase is, however, often far above the

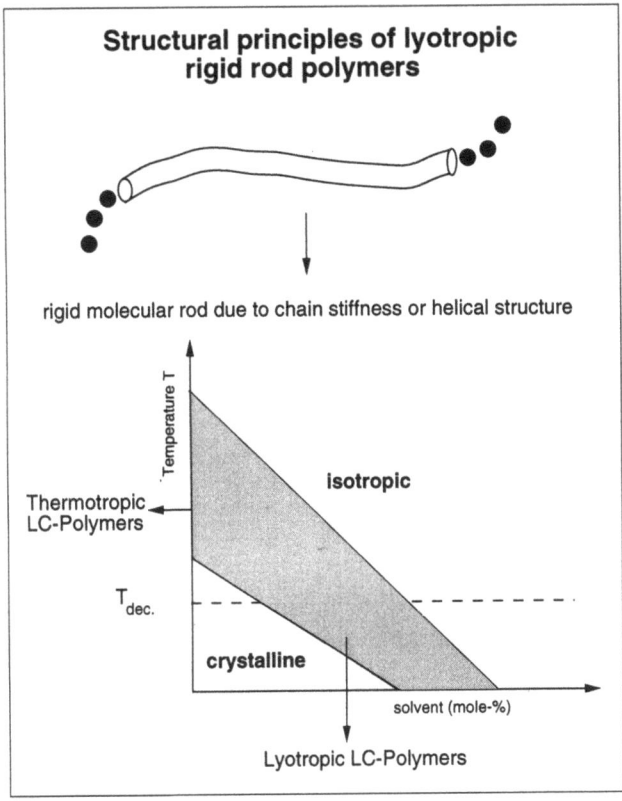

Fig. 3.4. Rigid rod polymers can give rise to thermotropic or lyotropic LC-phases depending on the decomposition temperature (T_{dec}).

decomposition temperature of the polymer. Therefore, the LC-phase in many polymers can only be observed after a solvent has been added. These phases are called "lyotropic" and the corresponding polymers "lyotropic" LC-polymers (see Fig. 3.4). The expression "lyotropic" – widely used throughout the literature – should however be used very carefully. For amphiphiles, for which lyotropic phases are well known (see Chapter 4, and [17]), the solvent leads to the formation of formanisotropic aggregates of the amphiphiles. These aggregates (e.g., cylinder micelles) are the constituent units of the lyotropic phases and they are only formed in the presence of solvents. The "lyotropic" phases of rigid rod polymers are however formed from the individual polymers, which organize above a critical concentration (volume ratio) into nematic LC-phases.

Figure 3.5 shows the structure of typical polymers which can form lyotropic LC-phases. The corresponding persistence length q, which can be used to calculate the relative degree of extension of the chains, is also shown.

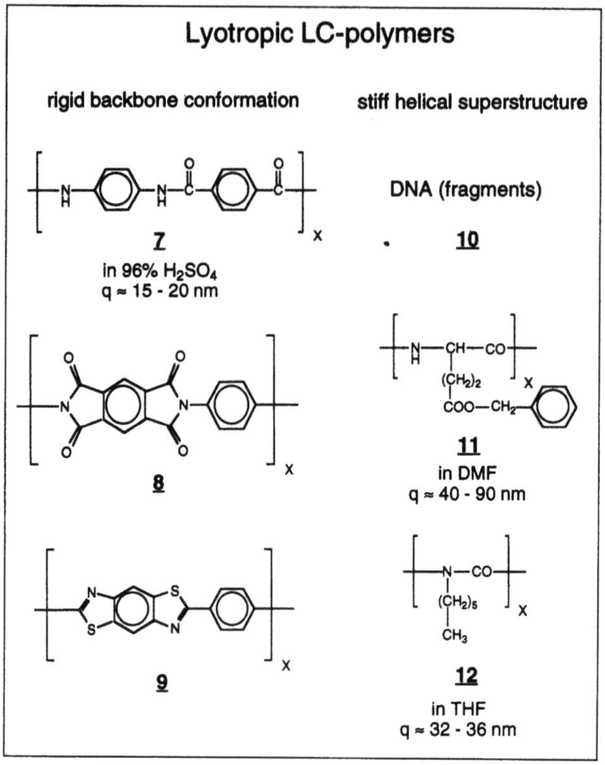

Fig. 3.5. Structures of polymers which form lyotropic LC-phases [9] and their persistence length q.

3.1.2.1 Theory

The theory for lyotropic LC-phases of polymers which resemble rigid molecular rods (rigid rod polymers) was mostly developed by Flory [1, 14]. Due to the predominant influence of the formanisotropic, stiff polymer chain, the theory of these systems does not start by considering primarily the anisotropic attractive forces between the polymer chains [18], as does the Maier-Saupe-theory or extensions of it (see Section 1.4.1.4). It starts rather by considering primarily the anisotropic repulsive forces (hard case repulsion) and includes the attractive forces later on [1, 19]. These theoretical studies predicted that rigid, rod-like polymers should undergo a spontaneous transformation from an isotropic to an anisotropic ordered solution above a threshold concentration, which in turn is an explicit function of the molecular axial ratio, x. A useful semi-empirical approximation for the critical volume fraction, v_p^*, for appearance of the anisotropic phase is:

$$v_p^* = (8/x)(1 - 2/x).$$

It was also shown by theory that at higher concentration, v_p^B, the isotropic phase becomes unstable and the solution becomes completely anisotropic. The basis predictions lead to the following relation:

$$v_p^B = 11.6/x.$$

Semirigid, or semiflexible polymers which, though highly extended compared to conventional random coiling polymers, nevertheless possess a significant flexibility, may also impart liquid crystallinity to their solutions. Calculations suggested that the theory for rigid rods may be adapted to semirigid chains by the simple device of replacing the molecular axial ratio, x, in the relation above by the axial ratio, x_k, of the Kuhn segment (the length of the Kuhn segment l_k is the length of one segment of the model chain and a measure of the chain stiffness):

$$x_k = l_k/d = (\langle r^2 \rangle_0/n)(\rho N_A/l_u M_u)^{1/2},$$

where l_k is the length of one segment in the model chain, d is the chain diameter, $\langle r^2 \rangle_0$ is the mean-square end-to-end length of the unperturbed, random chain, ρ is the density of the polymer, n is the number of repeat units in the real chain, l_u is the projected length of the repeat unit on the axis of the extended chain, and M_u is the molecular weight per repeat unit. Hence, there are two prerequisites for liquid crystal formation in polymer solutions: i) sufficient inherent rigidity of the molecular structure, and ii) sufficient solubility ($v_p > v_p^*$).

Poly-(benzyl-L-glutamate) (PBLG, polymer 11 from Fig. 3.5) has been most extensively studied. In all circumstances in which the liquid crystalline phase is present, PBLG is in the extended α-helical conformation. Hence, PBLG provides a perfect molecular analog of the rod-like particles considered in the theoretical modeling of phase transitions. The volume fraction v_p^* at incipient phase separation of the anisotropic phase was experimentally determined for different PBLG/solvent systems. The agreement between observed and calculated values of v_p^* is quite satisfactory.

The volume fraction v_p^*, at which the anisotropic phase first appears (Fig. 3.6, A points) and that at which the entire solution becomes a continuous birefringent liquid (B points) were determined optically for PBLG/dioxane systems with polypeptides of differing degrees of polymerization (since PBLG has a rather stiff helical conformation, an increase of the degree of polymerization leads to an increase of the axial ratio x). Subsequent study of shear viscosity for PBLG solutions as a function of polymer concentration has provided independent determinations of v_p^*. In Fig. 3.6 the calculated dependence of the A and B points on an axial ratio (from Flory's model) is shown with the experimental results. In general, the agreement between theory and experiments is rather good considering the idealizations incorporated into the former and the lack of sensitivity in the latter.

Measurements at higher temperatures show deviations from the original theory developed for real rigid rod. These deviations stem from changes of the stiffness of the polymer chain with temperature or from temperature dependent anisotropic dispersion forces.

Another point of interest is the miscibility of LC-polymers with each other and with amorphous polymers. This topic is important considering the miscibility rules of LC-phases and considering blends of rigid rod- and flexible polymers (molecular

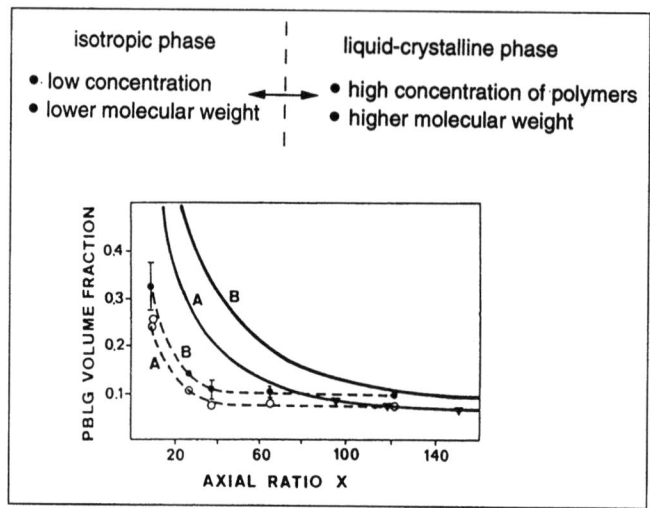

Fig. 3.6. Comparison of theoretical and experimental determination [9] of the incipient phase separation (A) and disappearance of isotropic phase (B). Solid curves: lattice theory, dashed curves: experiments.

reinforced polymers) [16]. In this context it is important to remember that the miscibility of polymers is very poor in general. Because of the high molar mass of polymers, very few centers of mass can be arranged independently. Therefore, the translational contribution to the entropy of mixing ΔS_{mix} (comparable to the entropy of mixing of an ideal gas) is very small. In addition, the change of intramolecular entropy is small, too, if two amorphous polymers mix. In order to meet the general thermodynamic requirement of a negative value of the Gibbs energy of mixing

$$\Delta G_{mix} = \Delta H_{mix} - T\Delta S_{mix} < 0,$$

a negative mixing enthalpy ($\Delta H_{mix} < 0$) is necessary for high polymers. This requires however that the interactions of polymer A with polymer B are more favorable than the interaction of both polymers with themselves. Miscibility in high molar mass polymers (including LC-polymers) is therefore very rare, even if both polymers possess the same LC-phase. In mixtures of rigid-rod polymers and flexible polymers (coiled polymers) another problem arises which is also purely entropic in origin. Since polymer coils hinder the parallel orientation of the rod-like polymers in the nematic phase, they are completely repelled from the nematic phase into the isotropic phase. Contrary to this, the presence of rigid-rod polymers does not hinder the arrangement of flexible polymers into a coil. Small amounts of rigid-rod polymers may therefore be present in the isotropic phase; this is, however, limited by the poor miscibility of polymers in general. This limits the possibility extremely to achieve molecular reinforced polymers [16].

3.1.2.2 Rheology

Since lyotropic LC-phases of rigid-rod polymers are used to make high modulus polymer fibers (see Section 3.1.2.3) and since rheology is most essential for the fiber spinning process, the rheology of these systems has been investigated in detail [20, 21]. Because of the high viscosity of these systems and of the difficulty to orient them in magnetic fields, all these measurements have however been made on unoriented samples, which change their orientation during the rheological experiments (for the viscosity of l.m.m. LC cf. Section 2.1.3).

One of the most striking phenomena observed on these, primarily unoriented samples, is the fact that the viscosity does not increase monotonically with increasing polymer concentration. As illustrated in Fig. 3.7 for poly-n-hexylisocyanate (polymer 12) in toluene, the viscosity as a function of the polymer concentration goes through

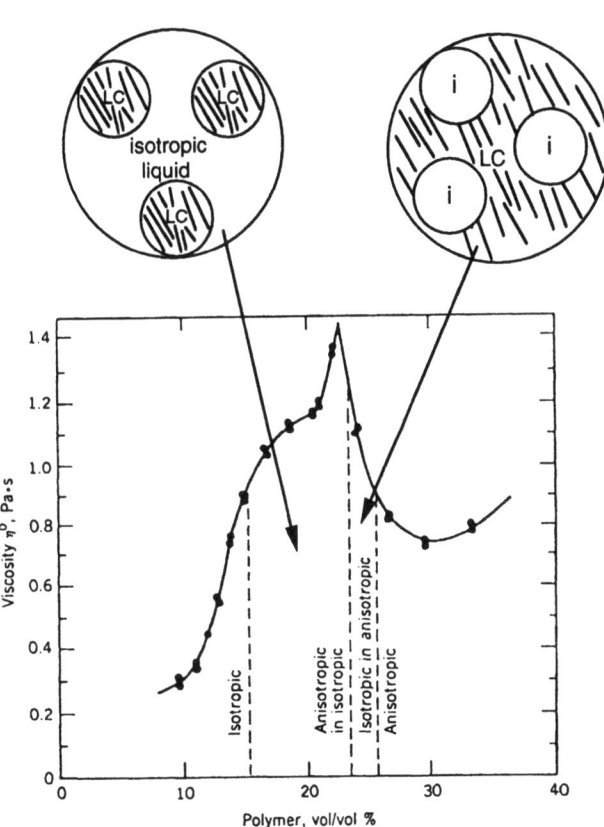

Fig. 3.7. Viscosity and phase transformation of poly(n-hexylisocyanate) (M_W: 73 000) in toluene as function of concentration [5].

a maximum which has been associated with the transition from the isotropic to the anisotropic phase. Later work, however, reveals that the first appearance of the anisotropic phase does not necessarily coincide with the maximum. Initial phase separation may occur at lower concentration, as shown in Fig. 3.7. Here, the highest viscosity is observed in the two-phase region. Despite the properties in the two-phase region, it is evident that the viscosity in the polymer-rich nematic phase can be lower than the viscosity of the isotropic phase, which is much poorer in polymers.

The same is true for a thermotropic transition from the nematic into the isotropic phase at higher temperatures. In this case a higher viscosity is measured for the high temperature isotropic phase than for the low temperature nematic phase. The reduction of the viscosity in the nematic phase is usually explained considering the anisotropic viscosity of LC-phases. In nematic phases of rigid-rod polymers great differences are expected for the different viscosity components (e.g., parallel and perpendicular to the director; cf. Section 2.1.3). A shearing of the sample will lead to an orientation of the (initially unoriented) sample, with the director parallel to the flow. In this case, mostly the lowest viscosity component is measured. In addition, the viscosity will be influenced by the movement and the rearrangement of disclination lines (cf. Section 1.4.1.1). The change of the orientation during flow also explains the increase of the viscosity at very low shear rates, which do not yet change the sample orientation. A quantitative understanding of the elemental processes governing the rheology has, however, not yet been achieved.

3.1.2.3 Application

Discussing the application of LC-polymers in fibers (like Kevlar) or as bulk materials, one has to keep in mind that they are used in a highly ordered (crystalline) state and not in the liquid crystalline phase. In this respect the importance of the LC-phase, induced by heating or by the addition of solvents, is limited to the processing step. Due to the low viscosities of lyotropic LC polymers as discussed above, the processing is easy and the orientation induced during flow (shear alignment) is the precondition to obtain a highly oriented crystalline material with excellent mechanical properties in fiber axis.

Theoretically, it is expected that the tensile modulus of a polymeric material composed of infinitely long fully extended chains is determined by the energies associated with the breaking or deformation of intramolecular bonds. The expected tensile strengths are very high. The presence of chain ends and especially the coiled chain conformation reduces this very high value to the much lower values observed for "normal" polymeric materials. Table 3.3 summarizes theoretical and observed tensile moduli of different polymeric materials. Whereas great differences are found for amorphous polymers, the moduli of the lyotropic LC-polymer $\underline{7}$ (see Fig. 3.5) are very close to the high theoretical value. This proves the extended chain configuration in the solid state. Taking into account the lower density of the polymeric materials compared to metals or glass, their specific tensile modulus (= modulus divided by density) becomes superior to inorganic materials (see Fig. 3.8).

Depending on the choice of processing conditions, fibers from polymer $\underline{7}$ (Fig. 3.5) can be produced from the lyotropic LC-phase with significantly different mechanical

Table 3.3. Comparison of calculated and experimental fiber moduli [9].

Polymer	Tensile modulus (g/d)	
	Calculated	Observed
cellulose	500-800	100-140
polyethylene	up to 4000	20-50
		after special processing
		up to 1400
nylon 66	1600	50
PET#	1200	110
polymer 7 (Fig. 3.5)	1500	1000 - 1400

#: Polyethylenterephtalate

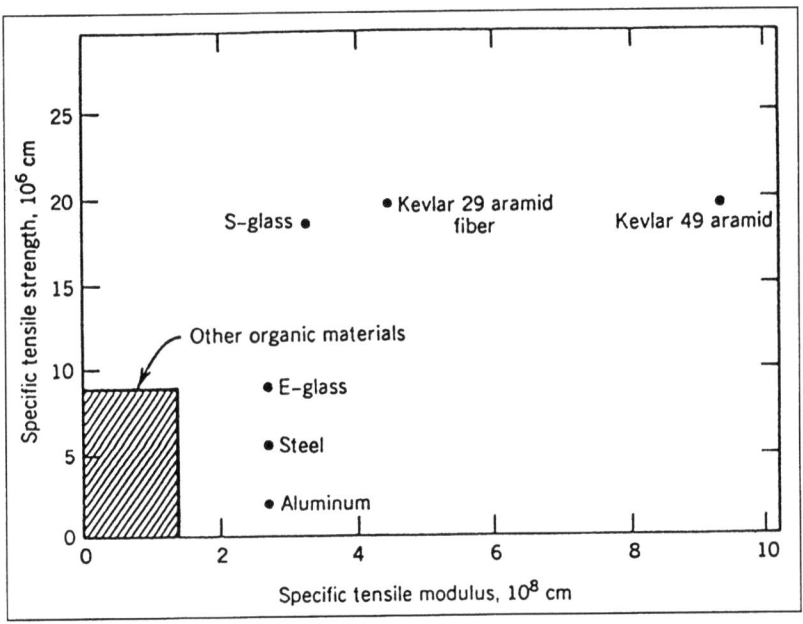

Fig. 3.8. Specific tensile strength (stress at break) and specific tensile modulus of polymer fibers and other materials [5] (tensile strength or tensile modulus divided by density).

properties. The best known commercial variants are Kevlar 29 and Kevlar 49 (Du Pont), whereby Kevlar 49 is heat-treated. Today, Kevlar fiber has replaced steel, fiberglass, asbestos, and graphite for a variety of applications including radial tires, armor-plates, brake linings, and composites. Kevlar 29 is more suited for use in areas where high impact resistance is of primary importance, such as in bullet-proof vest and cables of all types. The fiber is inherently flame resistant and begins to carbonize at about 427 °C. This makes it ideal for ventilation ducts, particularly those in mines. Kevlar 49 has been specifically designed for plastic reinforcement and is intended more for the aerospace industry where significant reduction in weight can be achieved over metallurgical products without compromising performance.

Concerning the thermotropic LC polymers, Vectra (Hoechst-Celanese), Xydar (Dartco), and Econol are commercialized. The low melt viscosity enables their use to injection-mold components with long or complex paths and thin sections, their utilization at very high filler loadings, and their application as processing aids. The coefficients of thermal expansion of these polymers are significantly lower than those of conventional isotropic polymers. As a consequence, the shrinkage is negligible in the mold, permitting precision molding of components. Their low thermal expansion coefficient approaches that of metals and glass. This enables their use in applications such as surface mount devices and optical fiber sheathing. In general, unfilled MC LCP's exhibit excellent mechanical properties. They have very high stiffness and strength, and are superior to glass fiber reinforced conventional thermoplastics.

3.2 Liquid Crystalline Side Chain Polymers with Calamitic Mesogens

3.2.1 Structures and Structure-Property Relations

The starting point for the rational synthesis of polymers with mesogenic side-groups was the "spacer concept," i.e., the idea to decouple mesogens and polymer chains by a flexible segment (the so-called spacer) [2,3]. In this way, two more or less independent subsystems are formed: the mesogens, which are free to form LC-phases and the polymer chains, which may adopt a coil conformation (see Fig. 3.9). Following this concept a wide variety of thermotropic LC-side chain polymers was synthesized [3, 5, 6, 9, 17, 22–24]. Due to their much lower viscosity compared to LC-main chain polymers, their liquid crystalline properties are rather similar to low molar mass (l.m.m.) liquid crystals and, due to the decoupling of the mesogens, similar structure-property relations are found.

Some of the molecular structures used to build up LC-side chain polymers are compiled in Fig. 3.10. Rather different polymer chains and mesogens are used. The spacer group consists in most cases of an n-alkyl chain, but oligo-ethylenoxide and oligosiloxane chains have been used as well. Newer developments include the use of vinyl ethers, which allow a living cationic polymerization [25] as well as H-bonded mesogens [26].

The structural variation of polymer subunits – mesogenic groups, spacer groups, and polymer chains – enables the influence of each of these constituents on the properties of the liquid crystalline polymers to be determined. The influence of variations of the mesogenic groups and the spacer length on the phase behavior is summarized in Table 3.4.

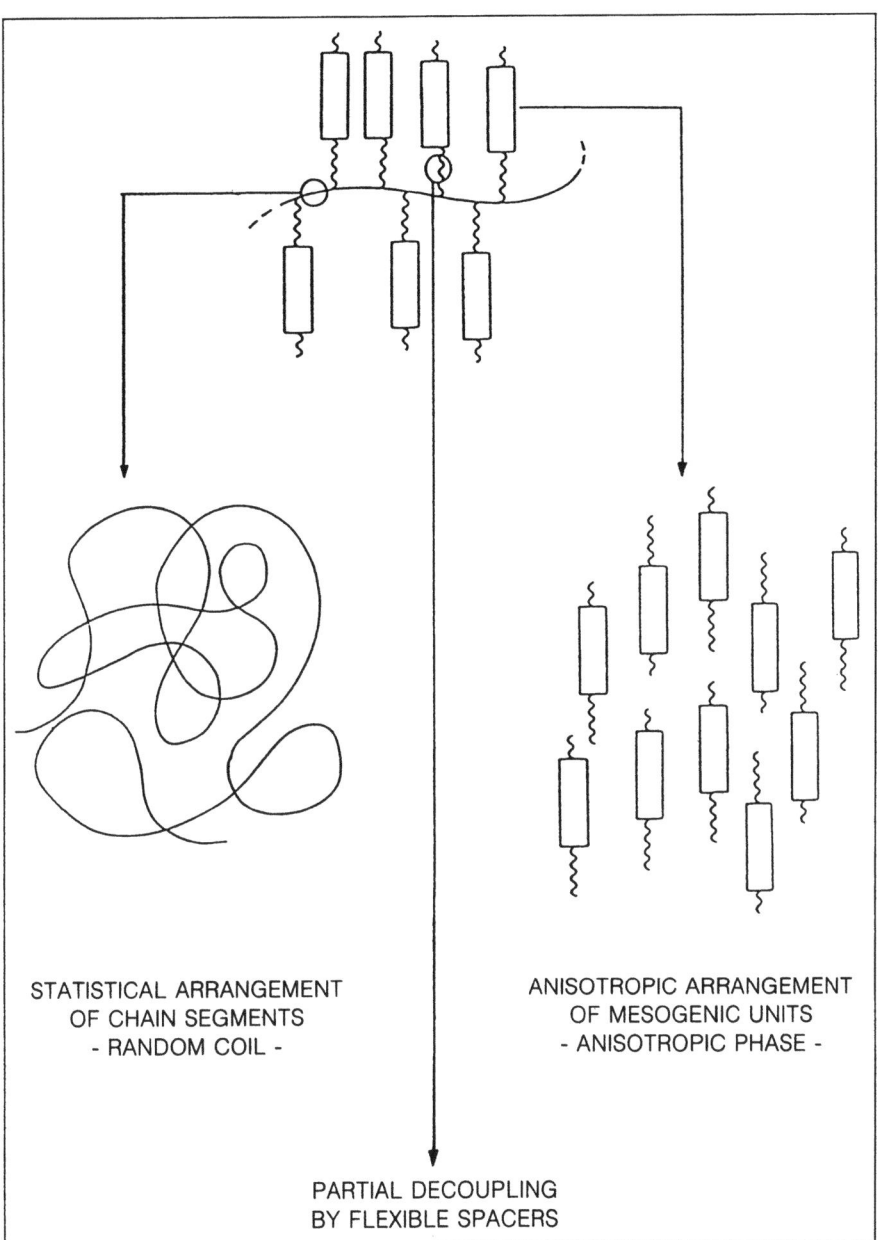

Fig. 3.9. The concept for the synthesis of LC-polymers: Mesogens with their tendency towards anisotropic arrangement and polymer chains with their tendency towards random coil formation are partially decoupled via flexible spacers.

117

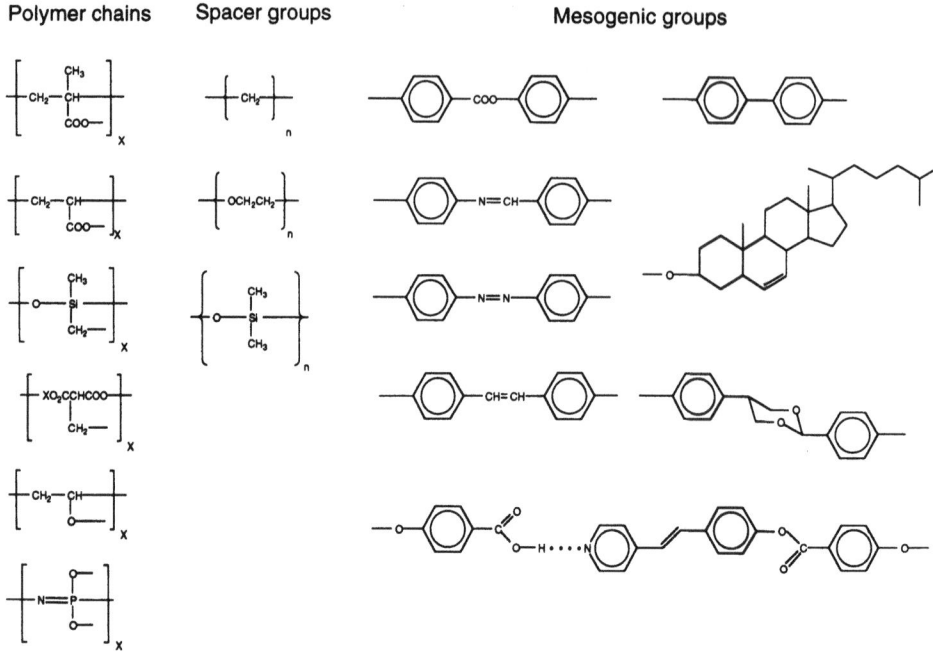

Fig. 3.10. Polymer chains, spacer groups and mesogens used to prepare LC-side chain polymers [23].

Table 3.4. The influence of variations of the mesogens and the spacer length on the phase behavior of the LC-side chain polymers **13–15** [24].

No.	n	R	Phase transitions [°C]
13	6	—⟨○⟩—COO—⟨○⟩—OCH₃	g 35 S_A 97 n 123 i
14	6	—CO—⟨○⟩—⟨○⟩—C₃H₇	g -10 n 20 i
15	2	—⟨○⟩—COO—⟨○⟩—OCH₃	g 62 n 116 i

The influence of varying the mesogenic structure [24] is demonstrated by comparing polymer 13 and 14. A polyacrylate with six methylene units as a spacer and benzoic acid phenyl ester as mesogen (polymer 13) shows a glass transition temperature at 35 °C, a smectic A phase up to 97 °C and a nematic phase that becomes isotropic at 123 °C. A change of the mesogenic group, in this case, the exchange of the benzoic acid phenylester with a phenyl-cyclohexane (polymer 14), keeping the rest of the polymer structure constant, leads to drastic changes in the phase behavior: the glass transition temperature drops to −10 °C and only a nematic phase is found for this polymer. This is in accordance to what is known from low molar mass liquid crystals: the phenylcyclohexanes are known for their low viscosities, low melting temperatures, and a preference for nematic phases.

The influence of a decrease in spacer length while keeping the mesogen constant is revealed by a comparison of polymer 13 and polymer 15. In this case, the glass transition temperature rises and only a nematic phase is observed for the polymer with the short spacer. This again is not astonishing because a decrease in the spacer length corresponds to a decrease in the length of the tails of the mesogens and it is well-known from low molar mass liquid crystals that this favors the formation of nematic phases (see Section 1.4.1.3). In conclusion, the main effect of varying the mesogenic groups and spacer length is both a change in the mesophase type and the clearing temperature as the upper limit of the liquid crystalline phase.

In contrast, the main influence of varying the polymer main chain [24] is summarized in Table 3.5. All three polymers in Table 3.5 show, independent of the polymer chain, a nematic phase. However, the glass transition temperatures are very

Table 3.5. The influence of variations of the polymer main chain on the phase behavior of the LC-side chain polymers 16–18 [24].

No.	M	n	Phase transitions [°C]
16	[CH$_2$–CH(CH$_3$)–COO]	2	g 97 n 120 i
17	[CH$_2$–CH–COO]	2	g 62 n 116 i
18	[O–Si(CH$_3$)–CH$_2$]	2	g 15 n 61 i

different. They are as high as 97 °C for the polymethycrylate 16, and only 15 °C for the polysiloxane 18. This certainly reflects the different mobilities of polymethacrylate, polyacrylate, and polysiloxane chains (T_g-values are 105 °C for PMMA, 10 °C for PMA, and − 120 °C for PDMS homopolymers, respectively). The fact that the glass transition of the polysiloxane 18 (15 °C) is still much higher than the glass transition temperature of the pure polymer backbone (− 120 °C) demonstrates that the mobilities are, of course, modified by the mesogenic groups. The influence of variations of the polymer chain on the clearing temperature are not so clear for the multitude of systems investigated. However, in most cases stiffer polymer chains (e.g., the polymethacrylate 16) lead to higher clearing temperatures than more flexible chains (e.g., polymer 18). Nevertheless, the main effect of a variation in the polymer main chains is a variation of the mobility and the glass transition temperature as the lower limit of the liquid crystalline phase.

3.2.2 Interaction of Polymer Chains and Mesogens

One of the major questions regarding the side groups polymers is: How effective is the decoupling of polymer main chains and mesogenic groups through the spacer? This problem shows up in the statistical-mechanical theory of LC-side chain polymers [27], which is more complicated than in the case of the main chain liquid crystals since there is a coupling of the alignment of the main and the side chains. The three possible cases are depicted in Fig. 3.11. In the case of the N_I phase the order parameter of the main chain is negative since it is aligned perpendicular to the nematic director. Phases N_{II} and N_{III} are more reminiscent of main chain LC-polymers since here the main chains are oriented along the preferred direction. Treating the interaction between the different parts of the polymer molecule in terms of a Maier-Saupe model, it is possible to discuss quantitatively the limits of stability of these phases.

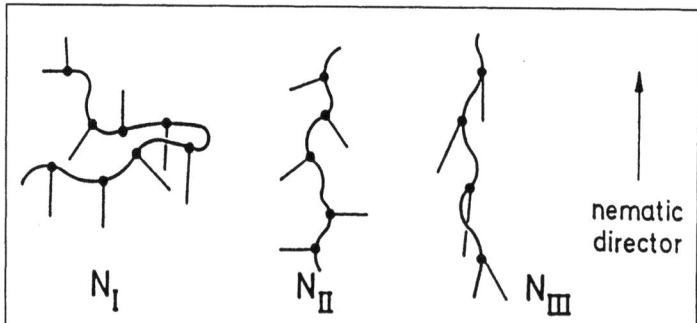

Fig. 3.11. Schematic representation of the three different possible arrangements of main chains and side groups in a nematic side chain polymer [27].

The model shown in Fig. 3.11 leads to the conclusion that the nematic order generally should lead to a distortion of the backbone conformation. This deformation is expected to be even more pronounced in the smectic state in which the main chains should be confined between the layers. Discussing the coupling of main chains and mesogens on a molecular level, it is important to differentiate between the coupling through chemical bonds (i.e., through the spacer) and through space by intermolecular interactions. Even for chemically "completely decoupled" systems (i.e., solutions of flexible polymer chains in low molar mass (l.m.m.) liquid crystals) some coupling exists through interactions through space. This leads, for example, to anisotropic chain conformations of amorphous polymers in liquid crystalline solvents. Investigations on these systems are however (due to a very poor solubility) limited to very low concentration.

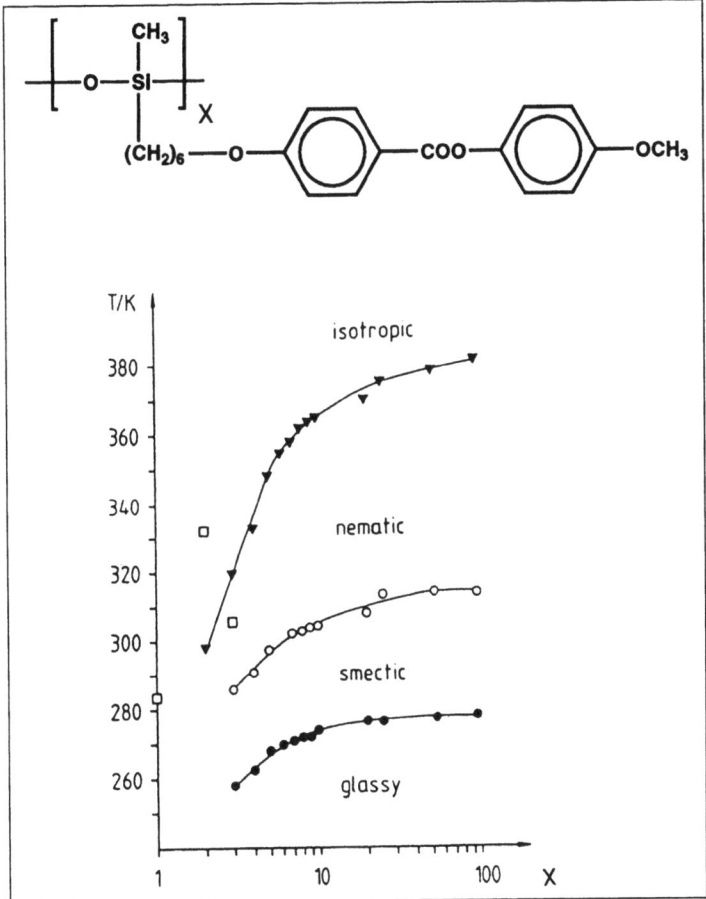

Fig. 3.12. Molecular weight dependence of the phase transition temperatures of a LC-side group polysiloxane [28] (x: degree of polymerization).

An early observation which proves the influence of the chemical linkage between main chains and mesogens on the properties of the LC system is the following: often monomers, which are formanisotropic but do not form thermotropic LC-phases themselves, give rise to thermotropic side chain polymers. Monomers with nematic phases give rise to smectic LC-polymers. This proves that the polymer fixation of mesogenic (formanisotropic) groups stabilizes LC-phases. The same behavior shows up clearly in the dependence of the phase transition temperatures (see Fig. 3.12) on the degree of polymerization. Whereas monomer and dimer do not form thermotropic LC-phases, the polymer has a smectic and a nematic phase over a broad temperature range. The width of the LC-phases increases strongly – starting with the trimer – and levels off for degrees of polymerization of more than 20. This behavior can be explained by the increase in density [28] upon polymerization or polymer fixation of the mesogens. The denser packing of the polymer fixed mesogens –compared to low molar mass (l.m.m.) liquid crystals – leads to a stabilization of the LC-phases, in good analogy to the effect of hydrostatic pressure in l.m.m. liquid crystals.

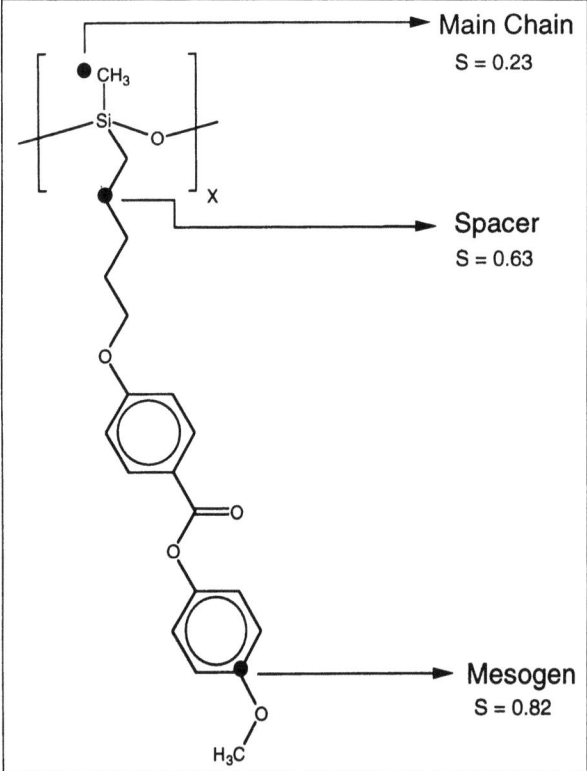

Fig. 3.13. Orientational order parameters of mesogen, spacer and polymer chain for an LC-side chain polymer [29].

The decoupling potential of the spacer for orientational correlations shows up in ^{2}H-NMR experiments [22, 29] (see Fig. 3.13). They show that the orientational order parameter S (cf. Chapter 1) is high for the mesogens (about comparable to l.m.m. liquid crystals), but they decrease rapidly along the spacer. Finally, the orientational order of the polymer chain is very low. Depending on the system, small positive or negative order parameters are found. This corresponds to a nearly perfect decoupling. These ^{2}H-NMR measurements prove also that all mesogenic groups take part equally in the LC ordering.

Neutron scattering offers the possibility to study the influence of the nematic or smectic order of the mesogens on the conformation of the polymer chains on a more global level [30, 31]. The experiments prove a small extension of the polymer chains parallel to the director in the nematic phase. In the smectic phase, however, the chains are confined between the layers (cf. also Section 3.2.3 and Figs. 3.16 and 3.20). If this interpretation is correct, then LC-side chain polymers would adopt about the same conformation as amorphous polymers dissolved in l.m.m. nematics. For these solutions, which represent the extreme case of perfect decoupling, the extension of the polymer chain parallel to the director is explained by interactions with the nematic field of the solvent. As a result, also neutron scattering shows (for the nematic phase) no evidence for additional orientational correlations between mesogens and polymer chains mediated by the spacer.

To summarize: The coupling between polymer chains and mesogens through a spacer is very important from the thermodynamic point of view, first to prevent a demixing and secondly to stabilize the LC-phases by a closer packing of the mesogens. It is presumably not important (at least for long spacers) for additional orientational correlations.

3.2.3 Influence of Additional Structural Variations on the Properties

Starting from the first (recognized as "classical") liquid crystalline LC-polymers (the side chain polymers **A** and the semiflexible main chain polymer **B**, shown in Fig. 3.14), the molecular architecture of liquid crystalline polymers has been varied extensively. This is illustrated schematically in Fig. 3.14.

In connection with the structural variation of side chain polymers, it was especially interesting to vary the number of mesogenic units per repeat unit in the polymer chain (**C** and **D** in Fig. 3.14). In order to increase the concentration of the mesogens, polysiloxanes substituted with alkylmalonic acid diesters ("dimesogenic" polysiloxanes, see **D** in Fig. 3.14) were utilized. In order to decrease the concentration of the mesogens [23, 24], copolysiloxanes were used in which only some of the siloxane units carried mesogenic groups. These copolymers have found interest for preparation of LC-polymers with low glass transition temperatures. If, in these copolymers, each mesogen is fixed separately to the polymer chain, then the liquid crystalline phase is lost at a ratio of 1 mesogen per more than 5 dimethylsiloxane units (see Fig. 3.15). If, however, "dimesogenic" polysiloxanes are used, then the ratio can be

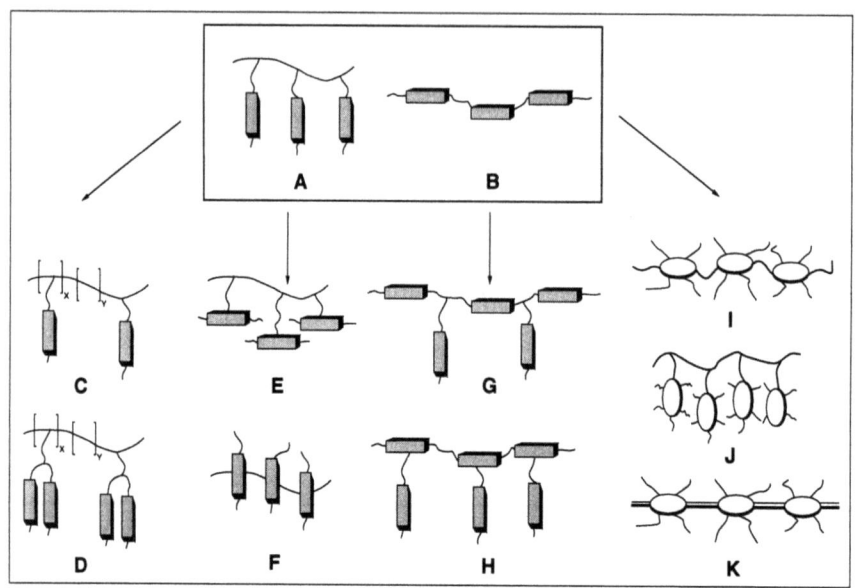

Fig. 3.14. Molecular architecture of LC-polymers: Variation of shape and arrangement of the mesogens.

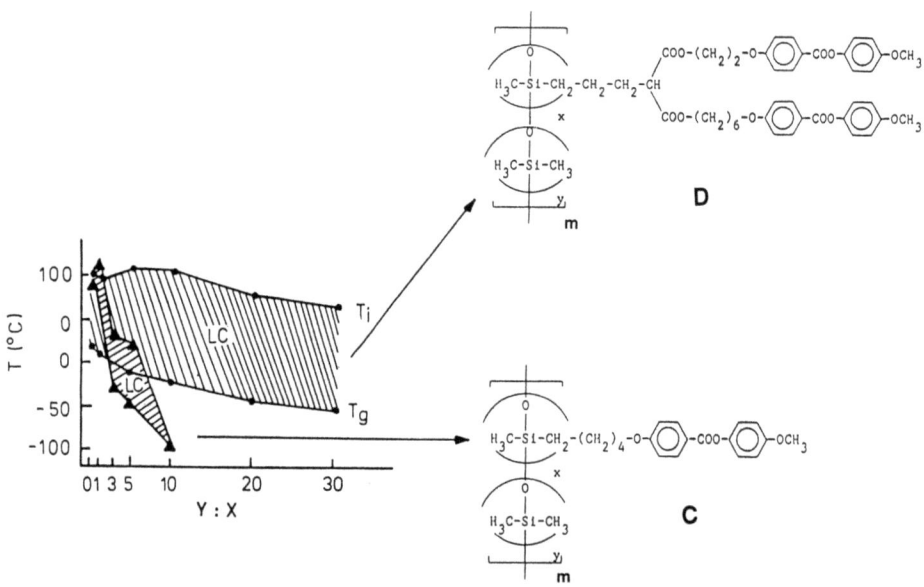

Fig. 3.15. Phase behavior of "monomesogenic" (type C) and "dimesogenic" copolymers (type D, Fig. 3.14) versus the ratio of repeating units carrying no mesogens (Y) to units carrying mesogens (X) [24].

as low as 2 mesogens (fixed together) per more than 30 dimethylsiloxane units (see Fig. 3.17). These copolymers show low glass transition temperatures and broad liquid crystalline phases, the phase width of which is comparable to the homopolymers.

The fact that "dimesogenic" polysiloxanes form liquid crystalline phases, even if only each 30th repeating unit is linked to mesogens, can be understood with a model derived from x-ray measurements [24] of homologous series of these polymers (see Fig. 3.16). The model assumes that mesogenic groups pack densely and reject the polysiloxane chains, which are incompatible with them. Thus, two sublayers are formed (microphase separation), one consisting of the disordered polysiloxane chains and the other of the mesogenic groups, which are in a smectic A arrangement.

However, if each mesogen is linked to the polymer chain separately, or if the polymer chains are more compatible (e.g., polyacrylates) with the mesogens, then the tendency towards microphase separation is reduced. In these cases, the incorporation

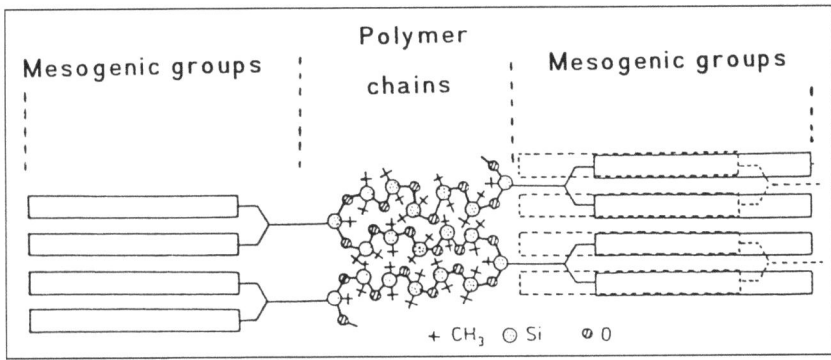

Fig. 3.16. Schematic model of the layer/structure for "dimesogenic" polysiloxanes (type **D**, Fig. 3.14) [24]. The mesogenic groups in dashed lines are positioned above the drawing plane.

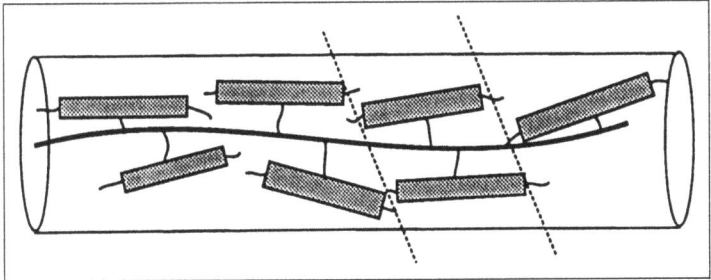

Fig. 3.17. Schematic drawing of the anisotropic polymer conformation [32] for LC-polymers with side-on-fixed mesogens (type **E**, Fig. 3.14).

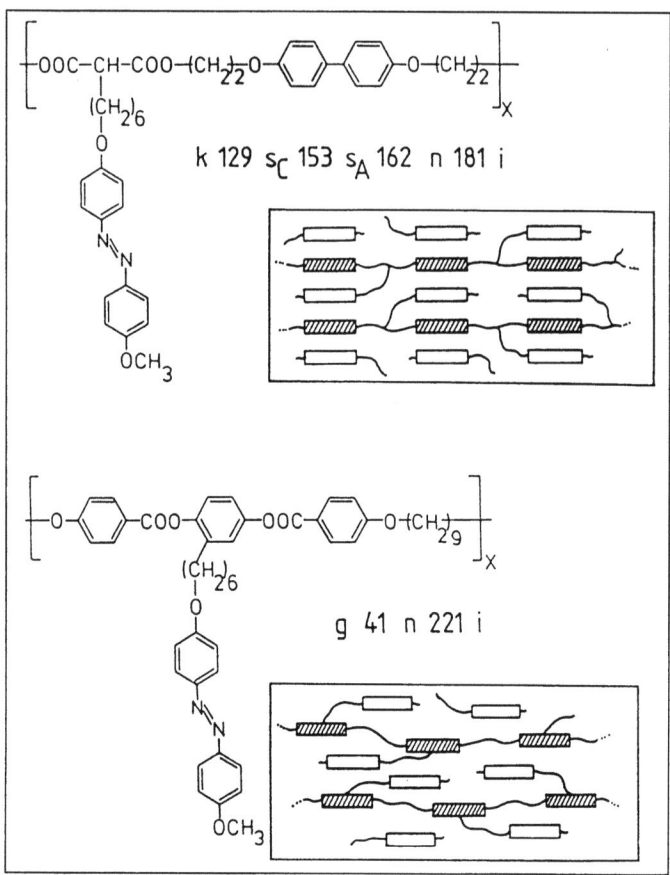

Fig. 3.18. Chemical structure, phase behavior and structural models of the mesophases of combined main chain/side chain polymers [24].

of nonmesogenic units into the polymer chain reduces the interaction among the mesogens and destroys the liquid crystalline phase.

A second variation of side group polymers led to side-on-fixed mesogen (type **E** in Fig. 3.14). This type of fixation preferably leads to nematic polymers and has two major consequences. Since a parallel orientation of polymer chains and mesogens is locally performed, a strong extension of the polymer chains parallel to the director results [32] (see Fig. 3.17). This strong deformation of the undisturbed coil conformation has led to the expression "jacketed" structure. It is different from the classical end-on-fixed mesogens (see Fig. 3.14, type **A**). Secondly, due to the hindered rotation of the side-on-fixed mesogens around their long axis, biaxial nematic phases result [33].

Another structural variation of LC-polymers led to the synthesis of combined main chain/side group polymers [23,24] (types **G** and **H** of Fig. 3.14) that combine the structural principles of the "classical" main chain and side chain polymers (types **A** and **B** of Fig. 3.14). These polymers show very broad liquid crystalline phases that are, in most cases, broader than the liquid crystalline phases of the corresponding pure main chain or side chain polymers. This suggests that the mesogenic groups in the main chain and the side groups are not oriented perpendicular to each other as suggested by the chemical formula in Fig. 3.18. They orient parallel to each other (see Fig. 3.18) to produce a uniaxial mesophase, as confirmed by x-ray measurements on oriented fibers. This structure, in which both types of mesogens are oriented parallel to each other, can also explain the differences in the phase behavior of polymers of types **G** and **H**. For polymers of type **G**, a smectic layer structure is performed locally and consequently mostly smectic phases are observed. For polymers of type **H**, however, the fixation of mesogenic side groups to the middle of the mesogenic groups in the main chain favors nematic phases (see Fig. 3.18).

3.2.4 Physical Properties of the Liquid Crystalline Phases

Due to the higher viscosity of LC-side chain polymers, compared to l.m.m. liquid crystals, their optical textures are atypical at first. However, on properly annealing

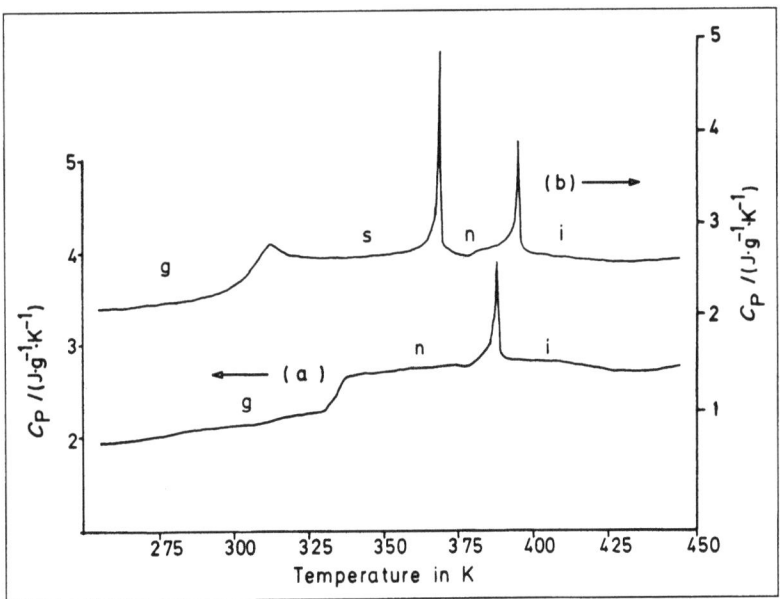

Fig. 3.19. DSC curve [50] of polymer 15 (a) and polymer 13 (b). (Dependence of C_p on temperature, heating rate 10 °C/min). g: glassy liquid crystal, S: smectic phase, n: nematic phase, I: isotropic melt.

the samples (sometimes up to several days) clear textures of nematic, cholesteric, and smectic A and C phases can be observed, which are quite analogous to l.m.m. liquid crystals (cf. Section 1.4.1.1). Differential scanning calorimetry (DSC) also shows comparable features (see Fig. 3.19). As a major difference to l.m.m. liquid crystals, mostly the absence of a melting transition is noticeable. Instead, a glass transition is observed which limits the LC-phases to lower temperatures. Below the glass transition temperature, the LC-structure and its orientation are frozen-in unchanged, but all mobility is lost (except for local jumping motions of parts of the molecule which do not change the overall shape of the molecules, e.g.: 180° flips of phenyl rings).

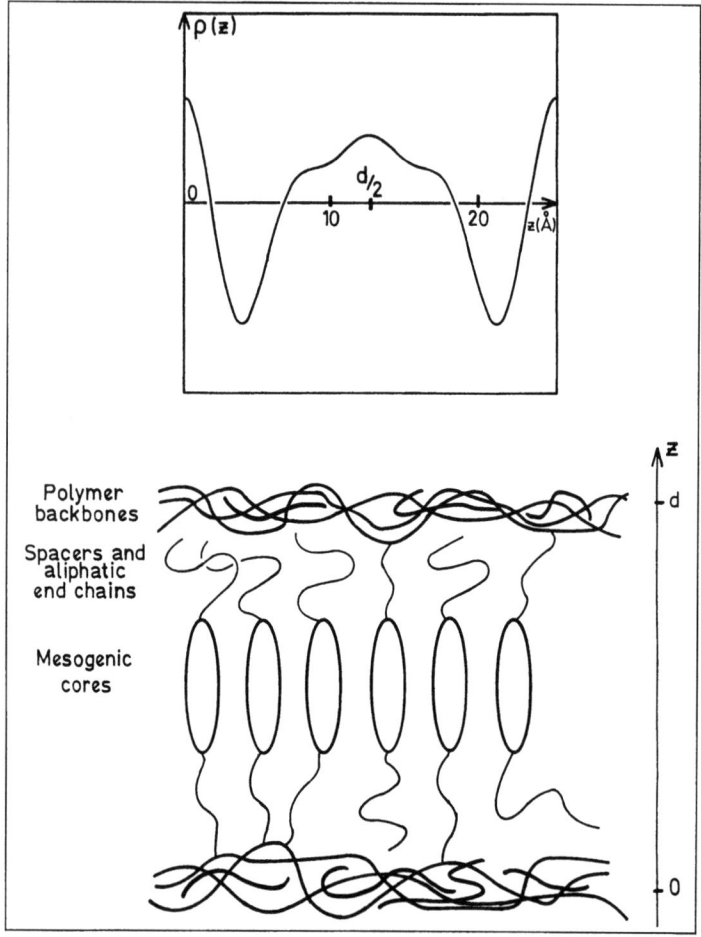

Fig. 3.20. A sketch of the molecular organization of LC-polymers in the smectic A phase [34] and a possible corresponding electron density profile.

X-ray measurements show diffractograms quite similar to l.m.m. liquid crystals which can be used for an assignment of nematic and different smectic phases in perfect analogy to l.m.m. liquid crystals. Differences are only found by detailed inspection [34] concerning two aspects:

- For smectic phases (even A or C) often up to five orders of the small-angle reflection (layer structure) are observed. This means that the density distribution of the smectic layer structure can no longer be simply sinusoidal and that more complex distributions and sharper interfaces have to be considered. Obviously, the polymer chains introduce additional correlations among the mesogens, which are linked to them. This effect has been especially investigated in polysiloxanes, for which main chains and mesogens are incompatible and tend to a confinement of the main chains at the interface of the smectic layers (Fig. 3.16 and 3.20).
- On overexposed x-ray films additional diffuse scattering shows up, which is due to uncorrelated longitudinal disorder and layer undulations.

Fig. 3.21. High-resolution electron micrograph showing undulated smectic planes and a dislocation [24, 35] for a combined main chain/side chain polymer.

Glassy frozen-in smectic phases offer some unique advantages for structural investigations. Because of their mechanical strength in the glassy state, thin films can be used for high-resolution electron microscopy. This makes a direct imaging of smectic layers [24, 35] possible (without replica and without staining), including the observation of single defects like edge disclinations (see Fig. 3.21).

Measurements of the orientational order parameter S by optical methods [17] and by ^{2}H-NMR [22, 29] in the nematic and smectic A phase reveal the following picture. At comparable reduced temperatures the order parameter is generally lower for LC-side chain polymers than for l.m.m. liquid crystals. ^{2}H-NMR spectroscopy proves, in addition, that all mesogens show the same order parameter. The results of these measurements are presented in Fig. 3.22. It shows that the increase of the order parameter with decreasing reduced temperature is slower for the polymeric systems compared to l.m.m. liquid crystals. However, since the temperature range of the polymeric LC-phases is often much broader, very high order parameters are found at

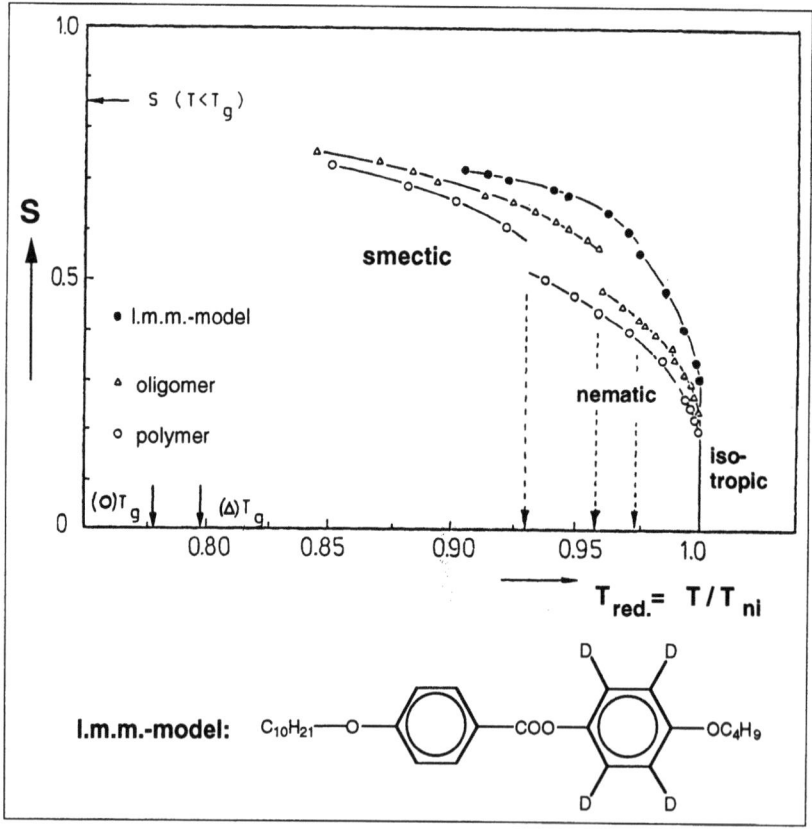

Fig. 3.22. Comparison of order parameters (^{2}H-NMR) for polymer <u>13</u> (as oligomer and high molar mass compound) and a corresponding l.m.m. model compound.

low temperatures in the glassy state. These maximal order parameters can be larger than for l.m.m. liquid crystals, in which crystallization destroys the LC-phase at low temperatures.

Director reorientations in electric fields [4, 22, 24] (like switching in LC-displays) of polymeric liquid crystals are possible in the nematic and chiral smectic C* phase (cf. Sections 2.7 and 2.8). They have been investigated for possible display applications (cf. Chapter 6). Concerning their dynamics, it has been found that they are about three orders of magnitude (factor 1000) slower than in l.m.m. liquid crystals and that they depend strongly on the molecular weight. Typical values are 200 ms to several seconds for nematics. For oligomers, the switching times can be comparable to l.m.m. compounds, whereas they reach infinity for LC-networks with infinite molecular weight. The molecular weight dependence of the switching time (director reorientations) can be explained by the anisotropic coil conformation of the polymer chain. Consequently, a reorientation of the mesogens (director) has to be followed by a reorientation of the polymer chains. This needs more and more time with increasing molecular weight and finally becomes impossible for LC-networks.

The threshold voltages of the Freedericksz-transitions in the nematic phase can be used to estimate elastic constants (cf. Section 2.7). It turns out that LC-side chain polymers with six or more methylene units as spacer possess threshold voltages and elastic constants comparable to l.m.m. liquid crystals. Shorter spacer groups however lead to an increase of threshold voltages and elastic constants. As an extreme, an increase of the elastic constants by a factor of about 100 has been observed on going from a spacer with six methylene units to a spacer with only two methylene units.

Properties of LC-side chain polymers and low molar mass liquid crystals differ mainly in dynamic aspects. This is demonstrated also in dielectric relaxation behavior (cf. Section 2.6.2). Dielectric relaxation measurements [24, 36] show both the dynamic glass process of the polymer chains (α-relaxation) and 180° jumps on the long axes of the mesogens (δ-relaxations; see Fig. 3.23). The δ-relaxation is orders of magnitude slower than in low molar mass liquid crystals and freeze at the glass transition temperature. In addition, up to three local relaxation processes are found below the glass transition temperature (see Fig. 3.23), which can be attributed to relaxations of the mesogenic groups (β), the spacer (γ_1), and the end group (γ_2). The results obtained for polymer 13 of Table 3.4 are presented in Fig. 3.23. It is interesting to note that the β-relaxation of the polymers 13 and 15 agrees completely with the dynamics of 180° phenyl flips determined for the same polymers by ^{2}H-NMR spectroscopy.

Among chiral LC-phases mainly the cholesteric and the chiral smectic C* phase [17, 22] have been investigated. The cholesteric phase was examined especially with respect to the selective reflection of light (cf. Section 4.2). Glassy frozen-in cholesteric phases are interesting as materials for circular polarizers and because of their reflection color.

Polymers with chiral smectic C* phases [37] have been especially examined with respect to their ferroelectric switching (see Sections 2.8 and 6.3). Since only polymers with long spacer were used for this purpose, the observed static properties are similar

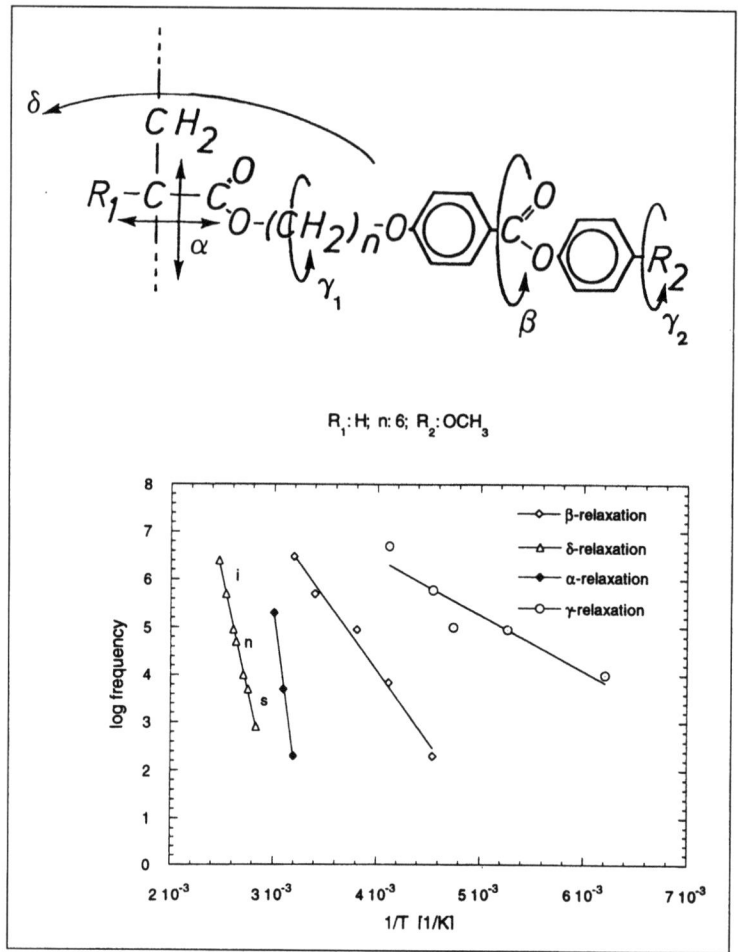

Fig. 3.23. Dielectric relaxation measurements of LC polyacrylates [24, 36]. Temperature dependence of the relaxation frequency for different relaxation processes for polymer 13 (Table 3.4).

to l.m.m. liquid crystals. The switching times are, however, again much longer (ms as compared to μs for l.m.m. liquid crystals).

As examples, the structure of some polysiloxanes with chiral smectic C* phases, their spontaneous polarization P_s and switching times τ are collected in Table 3.6 [37–39]. As in l.m.m. liquid crystals, the resulting polarization is not only a result of the magnitude of the lateral dipole moments. Steric factors – like the bulkiness – are effective as well.

Table 3.6. Structure, spontaneous polarization P_s and switching time τ of some ferroelectric polymers [37, 38].

Nr.	X:Y	—O—C(=O)—R*	Phase transitions [°C]	P_s [nC/cm^2]*	τ [ms]#
19	0:1	—O—C(=O)—CH(Cl)—CH(CH$_3$)—C$_2$H$_5$	k 68 S$_C$* 88 S$_A$ 156 i	130	—
20	0:1	—OC—(C$_6$H$_3$Cl)—O—CH(CH$_3$)—C$_6$H$_{13}$	k 62 S$_X$ 128 S$_C$* 170 i	140	—
21	0:1	—OC—(C$_6$H$_3$NO$_2$)—O—CH(CH$_3$)—C$_6$H$_{13}$	g 21 S$_X$ 57 S$_C$* 156 S$_A$ 183 i	211	17.3
22	3.2:1	—OC—(C$_6$H$_3$NO$_2$)—O—CH(CH$_3$)—C$_6$H$_{13}$	g 0 S$_X$ 46 S$_C$* 98 S$_A$ 144 i	104	3.9

*: values taken 10°C below the upper temperature limit of the S_C^*-phase
#: values taken 10°C below the upper temperature limit of the S_C^*-phase; applied field: 10 V_{pp}/μm

3.3 Liquid Crystalline Polymers with Disc-like Mesogens

Linking disc-like molecules through flexible spacers (types I and J, Fig. 3.14) leads to discotic LC-polymers [24, 29], i.e., polymers with discotic mesophases (cf. Section 1.4.2). If disc-like molecules are, however, linked very rigidly, different phases can result for which the term sanidic (and the symbol Σ from Greek: sanis = board) is used (cf. Section 1.4.3). The first discotic LC-polymers were side group systems incorporating hexapentyloxytriphenylene as the disc-like mesogen. More recently, some more polymers with disc-like mesogens have been synthesized. Structural variations, however, are still restricted and not to be compared to those of calamitic polymers. The disc-like mesogens so far incorporated are triphenylene ethers and benzene esters, all highly substituted with six long alkyl chains; and alkylated phthalocyanines (only rigid main chain polymers). The backbones are polysiloxanes, polymethacrylates, polyacrylates, and polymalonates for side chain polymers and polyphenylesters for main chain polymers. Some of the polymers prepared in this way are collected in Fig. 3.24.

Molecular engineering of discotic polymers has so far indicated some differences with respect to calamitic polymers. Polymerization generally increases or stabilizes

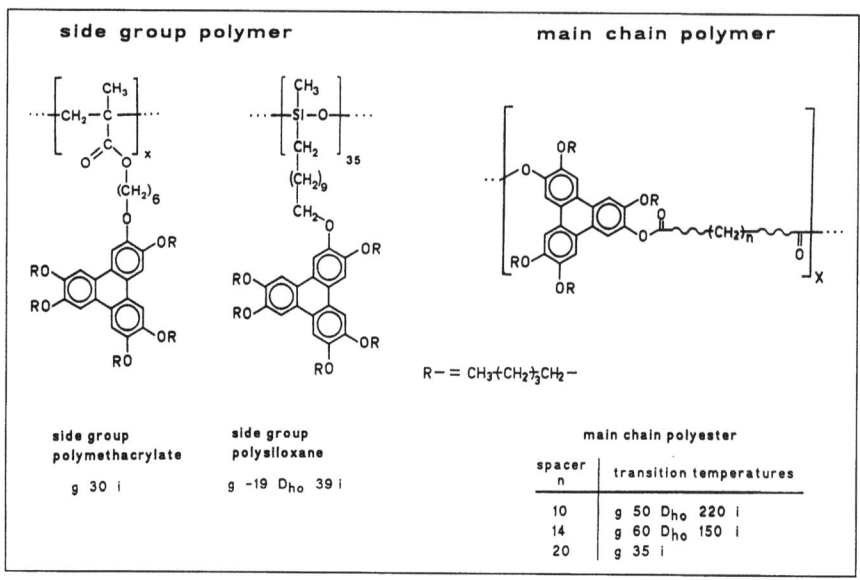

Fig. 3.24. Discotic LC polymers [24] with hexapentyloxytriphenylene-mesogens. For comparison: hexapentyloxytriphenylene (l.m.m. model compound): phase transitions (in °C): cryst. 69 D_{ho} 122 i.

the order for rod-like mesogens. This is not necessarily the case for disc-like mesogens: The discotic side chain polymethacrylate in Fig. 3.24, for instance, is not liquid crystalline and amorphous ($T_g = 30\,°C$); the preceding monomer, however, shows a highly ordered monotropic D_{ho}-mesophase (cf. Section 1.4.2.1). Likewise, the influence of the spacer in calamitic and discotic polymers is different. Long spacers do not prevent the formation of smectic mesophases for calamitic polymers, but lead to non liquid crystalline amorphous systems for discotic polymers. Thus, the discotic main chain polyester in Fig. 3.24 exhibits a D_{ho}-mesophase for spacer lengths $n = 10$ and $n = 14$, but is non liquid crystalline and amorphous for the higher spacer length $n = 20$. Recent experiments have shown that if the intracolumnar interactions are enhanced, such as through charge-transfer interactions, discotic polymers can stand more variations of molecular structure (longer spacers, bulkier polymer backbones) without losing the liquid crystalline phase. In this way, many amorphous discotic polymers with electron-rich triphenylene mesogens can be transformed into liquid crystals by adding electron acceptors: e.g., the amorphous side group polymethacrylate and the amorphous main chain polyester ($n = 20$) in Fig. 3.25 by adding 2,4,7-trinitrofluorenone (TNF).

To discuss a general relationship between polymeric structure and mesomorphic properties, the number of discotic polymers is far too small. For hexapentyloxytriphenylene containing polymers the following relationships are found. Polymerization or a change in polymeric structure do not bring about any change in the phase type, but influence the mesomorphic ranges, clearing temperatures and lower

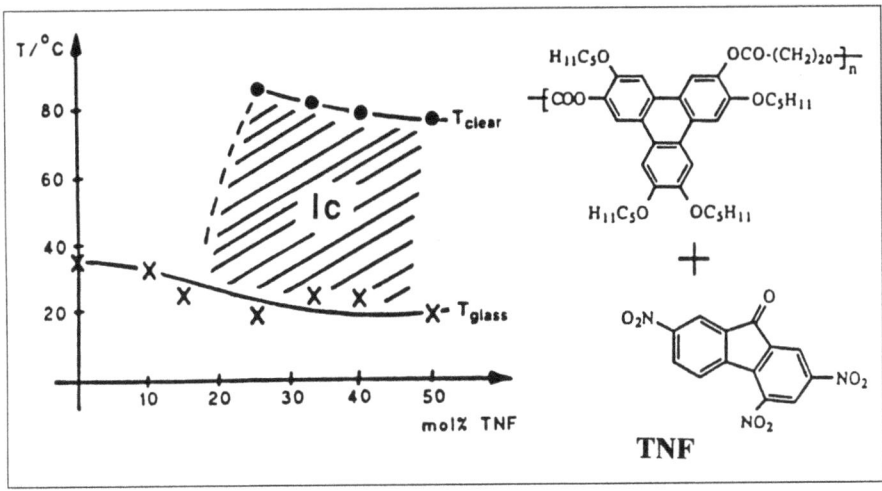

Fig. 3.25. Induction of a discotic LC phase by charge transfer interaction between an amorphous main chain polymer ($n = 20$, Fig. 3.24) and trinitrofluorenone (TNF).

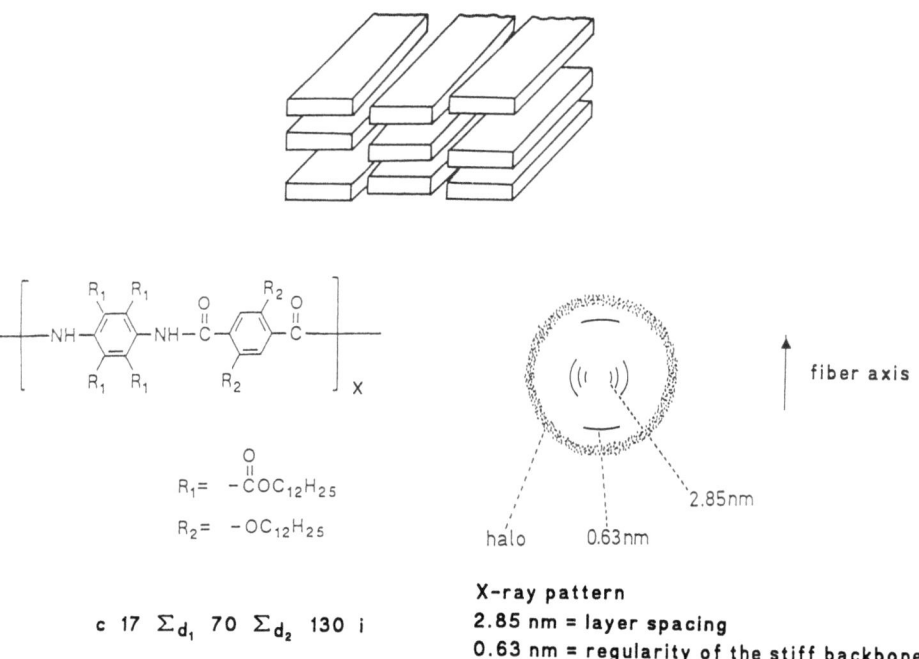

c 17 Σ_{d_1} 70 Σ_{d_2} 130 i

X-ray pattern
2.85 nm = layer spacing
0.63 nm = regularity of the stiff backbone

Fig. 3.26. Chemical structure, x-ray pattern and structural model [24] for a sanidic disordered phase (Σ_d) of a fully aromatic polyamide.

135

transition temperatures (melting point or glass-transition). The hexapentyloxytriphenylene, its side group polysiloxane, and its main chain polyester in Fig. 3.24 all exhibit one identical mesophase type. The temperature range of the D_{ho}-mesophase, however, varies from 50 °C for the hexapentyloxytriphenylene to 60 °C for the side group polysiloxane and 170 °C for the main chain polyester ($n = 10$). The clearing temperature is the lowest for the side group polysiloxane ($T_C = 39$ °C) and the highest for the main chain polyester ($T_C = 220$ °C).

Linking disc-like mesogens rigidly in the main chain leads to sanidic polymers, i.e., polymers with sanidic mesophases (see Section 1.4.3). The first sanidic polymer was a fully aromatic polyamide obtained through condensation of tetrasubstituted p-phenylendiamine and disubstituted terephthalic acid dichloride. The mesomorphic structure [40], as determined by x-ray investigations of oriented fibers, is neither calamitic nor discotic, but shows characteristics of both systems: the well defined layered structure of smectic (calamitic) phases and the one-dimensional packing in stacks of columnar (discotic) phases (see Fig. 3.26). The layer spacing (2,85 nm) corresponds to the maximum cross-section of the macromolecules perpendicular to their backbones.

3.4 Liquid Crystalline Elastomers

Very generally, LC-elastomers [22, 41–43] can be considered as being composed of two subsystems: the network of the polymer chains, which gives rise to rubber elasticity, and the mesogenic groups, which organize into LC-phases (see Fig. 3.27). Consequently, the great interest in LC-elastomers stems from the investigation of the interaction of both subsystems. At first, it might be considered that these interactions affect properties like LC-phases and phase transition temperatures, or the elasticity of the network. It has, however, been found that the order parameters and phase transition temperatures are nearly unaffected for low crosslinking densities. Thus, the crosslinking reaction, which transforms a soluble polymer into a soft solid (an elastomer), does not influence the LC order. The rubber-like elasticity of the elastomer is, on the other hand, preserved in the LC-phase.

The most fascinating aspect of the interaction of both subsystems is their reciprocal orientability. The use of differently structured LC-elastomers (see Fig. 3.28) is interesting in this respect, because the behavior of the two "subsystems" (network of polymer chains – mesogenic groups) should be different in each case. For crosslinked side group polymers there is only a weak coupling between the orientation of the mesogenic groups and the polymer chain, since they are decoupled by a flexible spacer. In crosslinked main chain polymers, however, the mesogenic groups are directly incorporated into the polymer chains. Crosslinked "combined" polymers may behave in an intermediate manner because one-half of the mesogenic groups is incorporated into the polymer chains, while the other half is linked to the polymer backbone as side groups via a flexible spacer. Both types of mesogens must, however, interact cooperatively to form an LC-phase.

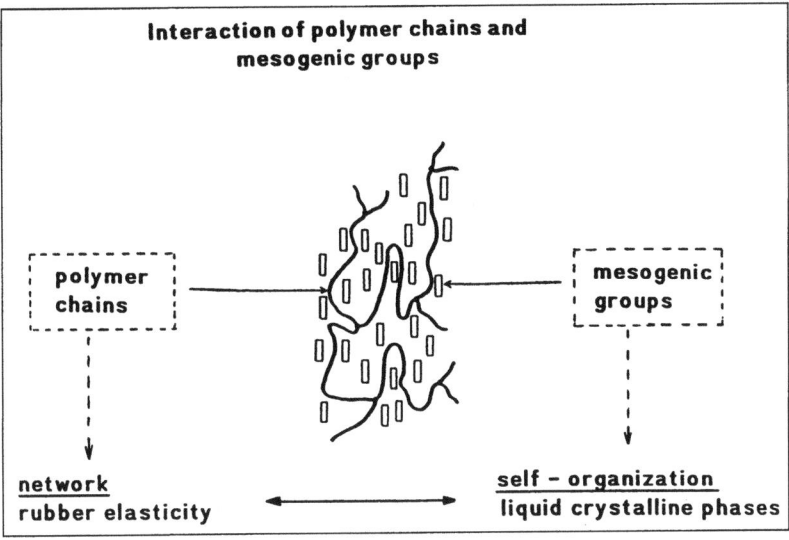

Fig. 3.27. Schematic representation of a crosslinked LC-elastomer [41]. The linkage between polymer chains and mesogenic groups is not yet specified.

The coupling of the orientation of the subsystems shows up in the mechanical orientability of the LC-phase. Small strains (about 20%) are enough to obtain an LC-monodomain in an LC-elastomer. Especially interesting is the possibility to shift the phase transition temperature (nematic-isotropic) by strain. Since mechanical stress induces a nematic-like order of the mesogens already in the isotropic phase, a shift of the phase transition temperature is expected [44]. For highly strained samples, the phase transition is expected to become second order and to disappear at last at a critical point. For real systems these investigations are limited due to tearing of the samples at higher strain. The maximal shift of the phase transition temperature observed so far is 0.8 K.

The interaction of the mechanical strain of the network with a helical superstructure can be investigated in LC-elastomers with cholesteric [45, 46] and chiral smectic C* phases [45, 47]. A concept has been presented to use this possibility for the preparation of new piezoelectric materials. It starts from the basic idea that an elastomer prepared from a ferroelectric material (chiral smectic C*) has to be piezoelectric. In this respect, it is the advantage of the elastomer that it can support strain under equilibrium conditions, whereas uncrosslinked LC-polymers and low molar mass materials are liquids, which relax shortly after a deformation. These piezoelectric properties [48, 49] could be proved recently for chiral LC-elastomers with cholesteric [45, 46] and smectic C* phases [45, 47]. As an example, the temperature dependence of the piezosignal of a smectic C* elastomer is given in Fig. 3.29 at a 10% sample compression.

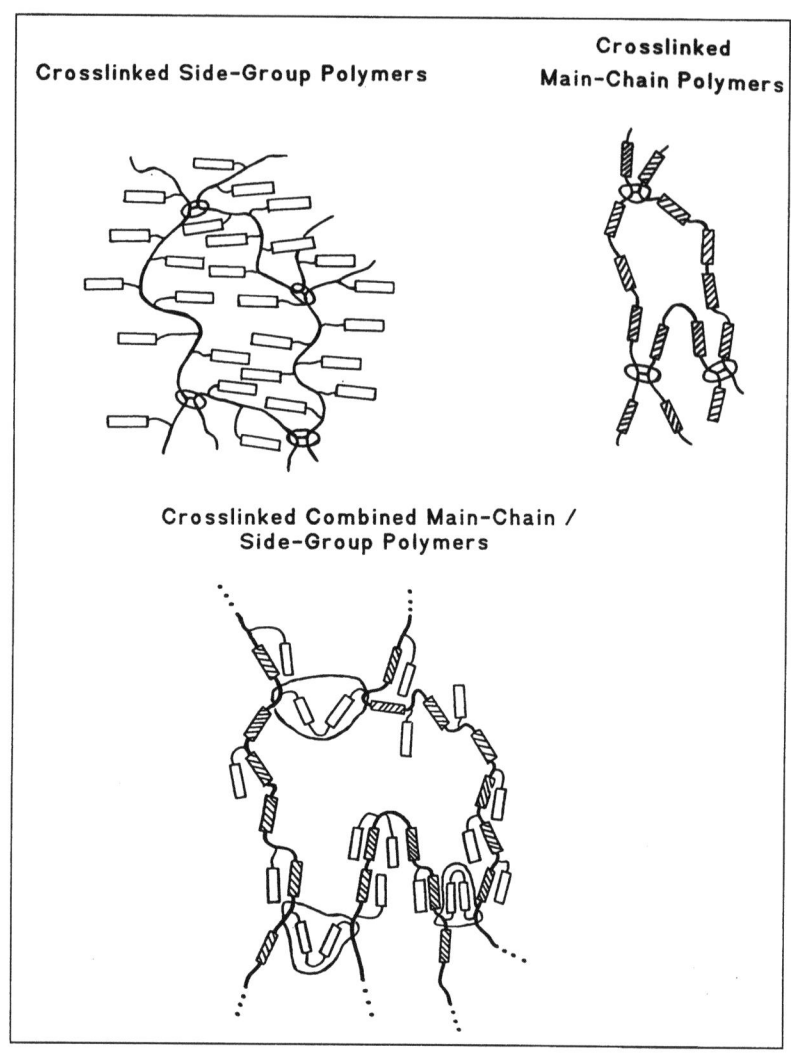

Fig. 3.28. Schematic representation [41] of different types of LC-elastomers.

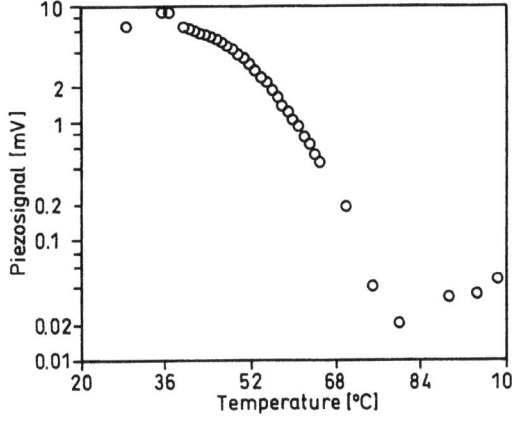

Fig. 3.29. Temperature dependence of the piezosignal (log) of a chiral smectic C* elastomer [45] at a sample compression of 10%. Phase transitions (in °C): g 1 S$_{C*}$ 81 iso.

References

1. Flory PJ (1956) Phase equilibria in solutions of rod-like particles. Proc R Soc Lond Ser 'A 234:73
2. Finkelmann H, Ringsdorf H, Wendorff JH (1978) Model considerations and examples of enantiotropic liquid crystalline polymers. Makromol Chem 179:2723
3. Shibaev VP, Platé NP, Freidzon YS (1979) Thermotropic liquid crystalline polymers, I: Cholesterol-containing polymers and copolymers. J Polym Sci Polym Chem Ed 17:1655
4. Platé NA (1984) Liquid Crystal Polymers I–III. In: Gordon M (ed) Advances in polymer science, Vol 59-61. Springer, Berlin
5. Kwolek SL, Morgan PW, Schaefgen JR (1987) Liquid crystalline polymers. In: Encycl Polym Sci, Vol 9:1, (ed.)
6. Ballauff M (1989) Kettensteife Polymere. Struktur, Phasenverhalten und Eigenschaften. Angew Chem 101:261
7. Weiss RA, Ober CK (1990) Liquid-Crystalline Polymers. American Chemical Society, Washington D.C., Vol 435
8. White JL (1985) Historical survey of polymer liquid crystals. J Appl Polym Sci Appl Polym Symp 41:3
9. Noël C, Navard P (1991) Liquid crystal polymers. Prog Polym Sci 16:55
10. Chiellini E, Lenz RW (1989) Main-chain mesogenic units. In: Allen G (ed) Comprehensive polymer science, Vol 5. Pergamon Press, Oxford, p 701
11. Blumstein A (1985) Nematic order in polyesters with flexible moieties in the main chain. Polymer J 17:277
12. Yoon D, Bruckner S, Volksen W, Scott JC (1985) Configurational characteristics and nematic order of semiflexible thermotropic polymers. Faraday Discuss Chem Soc 79:41
13. Yoon DY, Flory PJ (1989) Conformational rearrangement in the nematic phase of a polymer comprising rigid and flexible sequences in alternating succession. II. Theoretical results and comparison with experiments. Mat Res Soc Symp Proc 134:11
14. Furaya H, Abe A, Fuhrmann K, Ballauff M, Fischer EW (1991) SQID Studies of Main-Chain Polymer Liquid Crystals and a Rotational Isomeric State Treatment of the Data. Macromolecules 24:2999
15. D'Allest JF, Sixou JP, Blumstein A, Blumstein RB (1988) Small angle neutron scattering from liquid crystalline main chain polymers. Mol Cryst Liq Cryst 155:581

16. Ballauff M (1993) Structure of liquid crystalline polymers. In: Cahn RW, Haasen P and Kramer EL (ed) Materials Science and Technology, Vol 12. VCH, Weinheim, p 214
17. Finkelmann H, Rehage G (1984) Liquid Crystal Side Chain Polymers. In: Gordon M (ed) Advances in Polymer Science, Vol 60/61. Springer, Berlin, p 100
18. Luckhurst GR (1979) The physics of liquid crystals. Academic Press, London
19. Flory PJ, Ronca G (1979) Theory of systems of rodlike particles. Mol Cryst Liq Cryst 54:289
20. Wissbrun KF (1981) Rheology of rod-like polymers in the liquid crystalline state. J Rheology 25:619
21. Caffey CE, Porter RS (1984) Steady shear flow of solutions of rodlike macromolecules. J Rheology 28:249
22. McArdle CB (1989) Side Chain Liquid Crystal Polymers. Blackie and Son, Glasgow
23. Zentel R (1989) Polymers with Side-chain Mesogenic Units. In: Allen G (ed) Comprehensive polymer science, Vol 5. Pergamon Press, Oxford, p 723
24. Ringsdorf H, Voigt-Martin I, Wendorff J, Wüstefeld R, Zentel R (1991) Molecular Engineering of Liquid Crystalline Polymers. In: Fischer EW, Schulz RC and Sillescu H (ed) Chemistry and Physics of Macromolecules. VCH, Weinheim, p 21
25. Percec V, Tomazos D (1992) Molecular engineering of side-chain liquid-crystalline polymers by living cationic polymerization. Adv Mater 4:548
26. Kato T, Fréchet JMJ (1989) Stabilization of a liquid-crystalline phase through non-covalent interaction with a polymer side chain. Macromolecules 22:3813
27. Warner M (1989) The physical principles of side chain polymer liquid crystals. In: McArdle CB (ed) Side Chain Liquid Crystal Polymers. Blackie and Son, Glasgow, p 7
28. Stevens H, Rehage G, Finkelmann H (1984) Phase transformations of liquid crystalline side-chain oligomers. Macromolecules 17:851
29. Spiess HW (1993) Structure and dynamics of liquid-crystalline polymers with different molecular architectures from multidimensional NMR. Ber Bunsenges Phys Chem 97:1294
30. Davidson P, Noirez L, Cotton JP, Keller P (1991) Neutron scattering study and discussion of the backbone conformation in the nematic phase of a side chain polymer. Liq Cryst 10:111
31. Noirez L, Pépy G, Keller P and Benguigui L (1991) Smectic order and backbone anisotropy of a side-chain liquid crystalline polymer by Small-Angle Neutron Scattering. J Phys II France 1:821
32. Hardouin F, Mery S, Achard MF, Noirez L and Keller P (1991) Evidence for a jacketed nematic polymer. J Phys II 1:511
33. Hessel F, Herr R-P, Finkelmann H (1987) Synthesis and characterization of biaxial nematic side chain polymers with laterally attached mesogenic groups. Makromol Chem 188:1597
34. Davidson P, Levelut AM (1992) X-ray diffraction by mesomorphic comb-like polymers. Liq Cryst 11:469
35. Voigt-Martin IG, Durst H (1989) High-resolution images of defects in liquid crystalline polymers in the smectic and crystalline phases. Macromolecules 22:168
36. Zentel R, Strobl G, Ringsdorf H (1985) Dielectric relaxation of liquid crystalline polyacrylates and polymethacrylates. Macromolecules 18:960
37. Kremer F (1992) Special topic issue: Ferroelectric Liquid Crystal Polymers. In: Lewin M (ed) Polymers for advanced technologies, Vol 3. John Wiley & Sons, Chichester, p 194
38. Poths H, Schönfeld A, Zentel R, Kremer F, Siemensmeyer K (1992) Structure-Property Relationship Determining the Spontaneous Polarization in FLC-Polymers. Adv Mater 4:351
39. Kocot A, Wrzalik R, Vij JK, Zentel R (1994) Pyroelectric and electro-optical effects in the Sm C/ phase. J Appl Phys 75:728

40. Ebert M, Herrmann-Schönherr O, Wendorff JH (1990) Sanidics: A new class of mesophases, displayed by highly substituted rigid-rod polyesters and polyamides. Liq Cryst 7:63
41. Zentel R (1989) Liquid crystalline elastomers. Angew Chem Int Ed Engl Adv Mater 28:1407
42. Davis FJ (1993) Liquid-crystalline elastomers. J Mater Chem 3:551
43. Finkelmann H (1991) LC Elastomers. In: Ciferri A (ed) Liquid Crystallinity in Polymers. VCH, Weinheim, p 331
44. Schätzle J, Kaufhold W, Finkelmann H (1989) Nematic elastomers: The influence of external mechanical stress on the liquid-crystalline phase behaviour. Makromol Chem 190:3269
45. Valerien SU, Kremer F, Fischer EW, Kapitza H, Zentel R, Poths H (1990) Experimental proof of piezoelectricity in cholesteric and chiral smectic C/ phases of LC elastomers. Makromol Chem Rapid Commun 11:593
46. Meier W, Finkelmann H (1993) Piezoelectricity of cholesteric elastomers. Macromolecules 26:1811
47. Brehmer M, Zentel R, Wagenblast G, Siemensmeyer K (1994) Ferroelectric LC-Elastomers. Macromol Chem Phys 195, in press
48. Pleiner H, Brand M (1993) Piezoelectricity in cholesteric liquid crystalline structures. J Phys II 3:1397
49. Terentjev EM, Warner M (1994) Continuum Theory of Ferroelectric Smectic C/ Elastomers. J Phys II 4, in press
50. Portugall M, Ringsdorf H, Zentel R (1982) Synthesis and phase behaviour of liquid crystalline polyacrylates. Makromol Chem 183:2311

General

Ciferri A, Krigbaum WR, Meyer RB (eds) (1982) Polymer Liquid Crystals. Academic Press, New York.
Ciferri A (ed) (1991) Liquid Crystallinity in Polymers. Principles and Fundamental Properties. VCH, Weinheim
McArdle CB (ed) (1989) Side Chain Liquid Crystal Polymers. Blackie, Glasgow London

4 Lyotropic Liquid Crystals

K. Hiltrop

4.1 Introduction and Historical Background

Any discussion about "mesophases" and, especially, about "liquid crystals" might be thought of as dealing with only a marginal field of science. However, liquid crystals have in actuality become a consumer product with many everyday applications, substantiated by several examples in Chapter 6 as well as in Section 4.7 of this contribution. Additionally the variety of different structures exhibited by these phases of matter has attracted the attention of a diverse field of natural scientists and engineers, from both experimental and theoretical points of view.

Before going into detail the following simple definition of a "liquid crystal" is given and will be used throughout this chapter: "A liquid crystal is a mesophase which has partially or completely lost the long-range positional order of ordinary crystals, but still possesses one- or more-dimensional *long-range orientational* order of certain anisometric structure units" (which are defined in Section 4.2). In other words, the essential criterium for the liquid cristalline state is a certain fluidity and, at the same time, some uniformity (over "long" distances) of the average alignment of the building blocks of the phase.

Liquid crystals are usually divided into the two types "thermotropic" and "lyotropic" which may not always allow of an "either/or" distinction; many systems share characteristics of both. Both types possess a remarkable polymorphism and many different phases have to be distinguished. Thermotropic liquid crystals go through their phase sequence by adding or removing *heat* (Greek $\vartheta\varepsilon\rho\mu\eta$ = heat; $\tau\rho o\pi\eta$ = change) which is sketched in Fig. 4.1a.

The term "lyotropic" stems from the Greek word $\lambda v\eta\iota v$ = solve. Its meaning is that a lyotropic liquid crystal experiences its phase transitions by adding or removing a *solvent*.

Lyotropic mesomorphism can be visualized in the following two ways. From the first point of view, this state is formed out of a crystalline phase by the penetration of a liquid into the crystal lattice (see Fig. 4.1b). Mesophases between the anisotropic crystal and the isotropic liquid can then originate because the solvent must not break the crystal into *monomers* in any case, which would result in an isotropic solution; in a first step, the solvent can leave distinct intermolecular arrangements of the crystal quasi unchanged. Thus, it indeed enhances the mobility of *fragments* of the crystal and endows the system with liquid-like properties. But inspite of the fluidity of the new phase a certain *orientational order* can be maintained on a long-range scale. By increasing the solvent concentration the shape and size of the intermolecular arrangements can be modified due to a complex interplay of forces discussed in more detail in Sections 4.2 and 4.6, giving rise to the polymorphism of the mesophases.

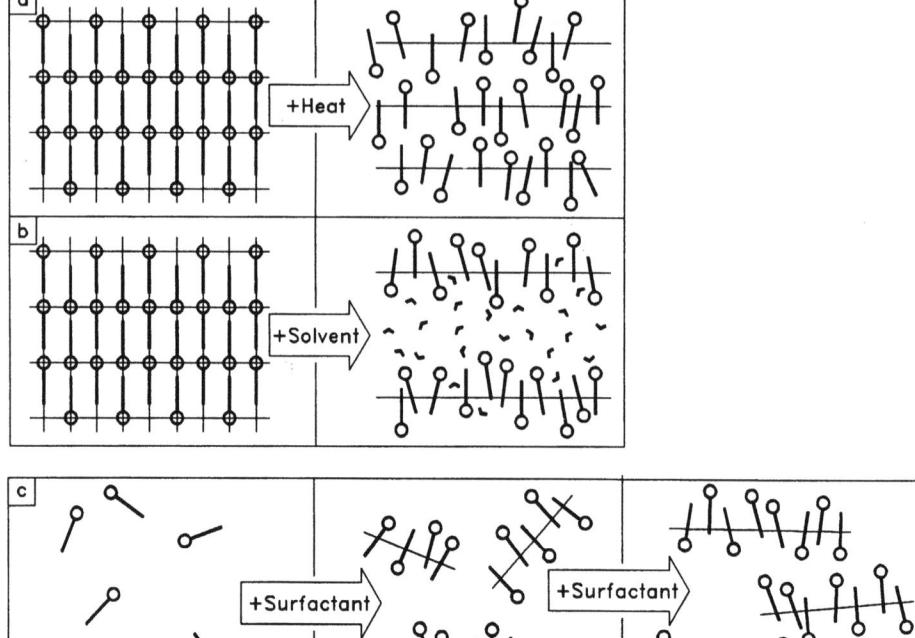

Fig. 4.1. Types of liquid crystalline mesomorphism.
a) Transition of a crystalline solid to a thermotropic smectic liquid crystal by the action of heat.
b) Transition of a crystalline solid to a lyotropic lamellar liquid crystal by the addition of a solvent.
In case a) and b) the long-range positional order breaks down completely in two dimensions (within the layers), while a long-range orientational order is maintained to some degree of order.
c) Formation of a lyotropic liquid crystal by an orientational ordering of anisometric solute aggregates due to increasing solute concentration.

In the second approach, lyotropic mesomorphism may be seen conversely starting from a pure solvent which is an isotropic liquid. Adding a suitable solute can result in the formation of solute *aggregates*, which is explicitly discussed in Section 4.2. Increased solute concentration may lead to an *anisometry* of the aggregates which at still higher concentrations can develop an inter-aggregate long-range orientational order; thus, a liquid crystalline state of the solution is reached. This way of approaching the liquid crystalline state is pictured in Fig. 4.1c.

To date, the standard solvent is water. Typical solute materials showing lyotropic mesophases with water are soaps and other ionic and nonionic compounds. They may be monomeric as well as polymeric. From a phenomenological characterization, one kind of suitable solute molecules possess a common feature that is called "surface activity". These "surfactants" prefer locations in water/air or water/oil interfaces. The creation of large oil/water interface areas plays a crucial role in the formation of lyotropic mesophases, as is explained in Section 4.2. It must be emphasized that surface activity of the solute is not an absolutely necessary precondition for lyotropic mesomorphism, as will be mentioned to some extent in Section 4.2.2. Nevertheless, all of Chapter 4 will mainly refer to aqueous lyotropic systems built up by surfactants.

About 3000 years A.D. the Sumerians boiled various oils and alkali together to prepare detergent solutions, which they used for washing their textiles as well as for medical recipes [1]. In the 4th century the word "saponarius" was recorded, which means "soap boiler". In the 19th century soaps became an industrial product. Therefore, it must be supposed that man came in contact with lyotropic liquid crystals a very long time ago, although presumably no details about this matter have been known at that time; however, in that respect, lyotropics are much older than thermotropics, the 100th anniversary of which was celebrated in 1988. The soap boilers were presumably the first who recognized qualitative differences between isotropic and liquid crystalline, as well as between two lyotropic liquid crystalline phases due to different viscosities. They coined the expressions "middle" and "neat" phase which are synonymous with the actual and more pregnant terms "hexagonal" and "lamellar" phase. One or both of these two liquid crystalline phases appear in most of the phase diagrams of lyotropic systems.

4.2 Hydrophobic Effect and Aggregation

In this Section a precondition for the occurrence of lyotropic liquid crystallinity is discussed, namely, the existence of anisometric structure-building elements. Referring to this point, first the crucial role of the solvent will be illuminated.

4.2.1 The Hydrophobic Effect

Water is a rather exceptional solvent. Besides its strong polarity, one important feature of its structure is the capability to form inter-molecular hydrogen bonds. This leads to a tetrahedral lattice of crystalline water with a relatively low density and, therefore, much free volume between neighboring water molecules. The ice structure is partly retained in liquid water. A large number of allowed positions of the hydrogen atoms is one consequence of the tetrahedral surrounding maintained by hydrogen bonds. Thus, the entropy of the phase is especially large. In order to solve nonpolar molecules the water structure can rearrange to form cages enclosing the solute; therewith the water gets more ordered. But for large solute molecules the bridges between adjacent water molecules have to be broken. Therefore, if the solved molecule is unable to form hydrogen bonds with water, the solving process is quite entropy-costly, and, consequently not favored if the entropy of mixing does not overcompensate it. This so-called "hydrophobic effect" is the dominant reason for the scarce solubility of many hydrocarbons in water [2].

4.2.2 The Aggregation

As an example, the solubilities of the following similar molecules are compared: a non-amphiphilic molecule (e.g., decane), an amphiphilic one with a weakly hydrophilic part (e.g., decanol), and an amphiphilic one with a strongly hydrophilic part (e.g., sodiumdecylsulfate). The solubility of decane or decanol in water is almost zero. On the other hand, sodiumdecylsulfate mixes with water over a wide range of concentrations. This is explained by the ability of the latter type of molecules to build up *aggregates*, which is a consequence of their balanced amphiphilic properties. Amphiphilic molecules are composed of hydrophobic (lipophilic) as well as of hydrophilic (lipophobic) groups (Greek $\lambda\iota\pi o\varsigma$ = fat; $\varphi o\beta o\varsigma$ = fear; $\alpha\mu\varphi\iota$ = both; $\varphi\iota\lambda o\varsigma$ = friend). Examples are given in Fig. 4.2. The hydrophilic part which can be ionic or nonionic, is usually called the "headgroup", while the hydrophobic one is called the "tail". Thus, there are two parts with extremely different solubility properties within one molecule.

By the aggregation both of the two opposite parts of the amphiphilic molecules profit by gathering into locally hydrophobic respectively hydrophilic micro-domains. Thus, a vast amount of internal surfaces, which are boundaries between hydrophilic and hydrophobic regions, is created. Nevertheless, there is no real micro-phase separation; rather, the phase as a whole must be regarded as remaining homogeneous (see Section 4.4).

Small aggregates with "finite" size (typically, they consist of 50 up to several hundred monomers) are called *micelles*. Rough drafts of micellar aggregates are sketched in Fig. 4.3. It must be emphasized that these drawings lack many features of real aggregates, regarding, for example, the regularity of their shape, which is grossly exaggerated.

In the previous literature the term "micelles" often does not refer to large (infinite) aggregates which were assumed to be essential for the liquid crystalline phases (see Section 4.3). However, it is doubtful if this discrimination is reasonable for real systems; rather, there is experimental evidence for only moderately large aggregates (micelles) also in liquid crystalline phases [3].

Micelles do not exist below a so called c̲ritical m̲icelle (formation) c̲oncentration, the "cmc". In less concentrated solutions only monomers and very small pre-micellar aggregates (mainly dimers and trimers) of the amphiphile occur. The setup of a micelle is a cooperative phenomenon; it takes at least a minimum number of monomers (about 50) to sufficiently reduce the water/hydrocarbon contact by the aggregation. Thus, it is plausible that no aggregates with intermediate sizes can be found. The cmc is usually temperature-dependent.

After exceeding the cmc, the amount of the non-micellar solved amphiphile changes only slightly with concentration. Instead, the excess surfactant forms more

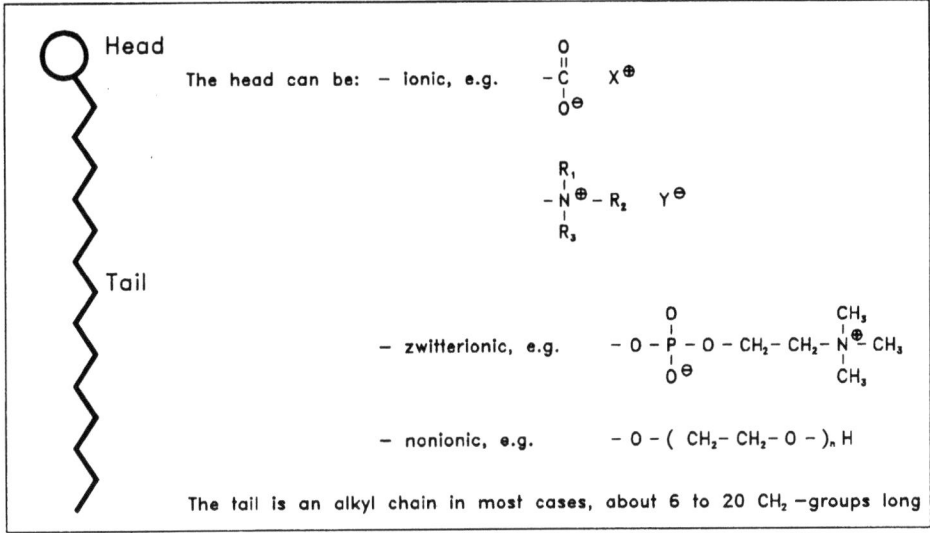

Fig. 4.2. Examples for amphiphilic molecules, composed of a polar hydrophilic headgroup and a hydrophobic tail.

aggregates and, thus, the *number* of micelles grows. There is an equilibrium between micellar and monomeric surfactant. The "aggregation number", i.e., the number of monomers per micelle, is more or less polydisperse, depending on the shape of the micelle and on thermodynamic parameters of the system. Spherical micelles do allow only a narrow size distribution because of the obvious geometrical restrictions. In some cases a second and even a third "critical" concentration ("cc2" and "cc3") may be crossed, at which further changes of physical properties occur and the micelles become *anisometric* (see Fig. 4.3b, c) [4].

Rod-like (prolate) as well as disc-like (oblate) micelles can develop, depending mainly on surfactant properties, like attractive and repulsive forces and the molecular geometry. A summary of the acting forces and of the geometrical aspects is given in Section 4.6.

In "normal" micelles the polar headgroups of the internal surfaces separate the hydrophobic core of the aggregate from the hydrophilic surrounding. If the amphiphile concentration is increased more and more, the water content of the solution may become too small to act as the continuous medium of the phase, i.e., to separate discrete micelles. Instead, the system turns over to "reverse" or "inverted" micelles; the curvature of the polar/unpolar boundary layer of the aggregates changes its sign. The hydrophilic solvent now represents the core of the micelle, and the hydrophobic chains and eventually added oil take the part of the continuous medium between the reverse micelles (see Fig. 4.3d).

Below a well-defined temperature the crystalline amphiphile is almost insoluble in water: no micelles can form and only hydrated crystalline modifications of the surfactant exist. This solubility temperature is concentration dependent.

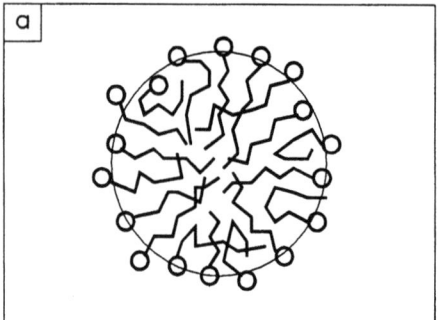
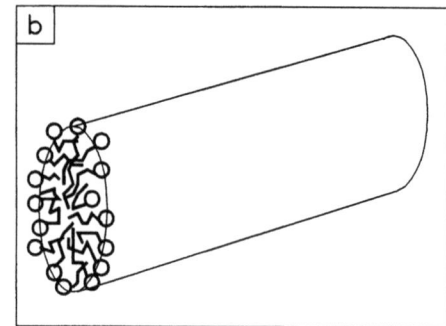
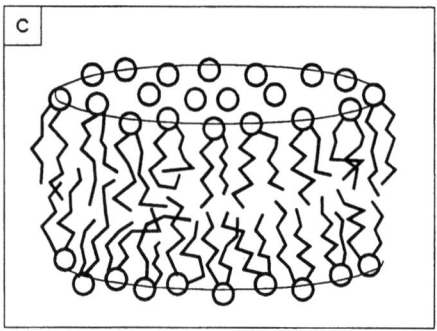
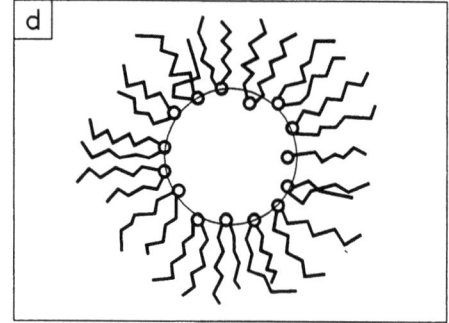

Fig. 4.3. Simplified micelle models; in normal micelles the hydrophilic headgroups form a shell between the hydrophilic solvent and the hydrophobic core of the aggregate, the tails are in a liquid-like state.
a) Spherical micelle;
b) Rod-like micelle;
c) Disc-like micelle;
d) Reverse spherical micelle; here, the hydrophilic solvent is inside the aggregate. The tails and eventually present oil make up the continuous medium.

A "Krafft-point" T_K is defined by the intersection of the cmc and the solubility curves in temperature versus concentration diagrams (see Fig. 4.4). Typical values of T_K are below resp. near room temperature for many surfactants. T_K increases with the aliphatic chain length. Only above T_K (more precisely: above the solubility temperature, which in the literature is also occasionally termed T_K) a sufficient amount of free monomers can exist in the solvent to exceed the cmc.

A necessary condition for a surfactant molecule to fit into a micelle is the "chain melting"; an aliphatic chain has to be flexible to accept suitable conformations for filling up the space in the micellar core. (The chain melting may occur when the solubility curve of a surfactant is crossed; however, these two phenomena must not coincide.)

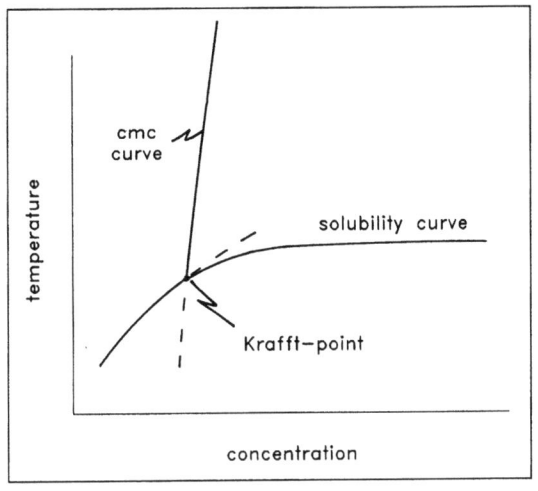

Fig. 4.4. Critical micelle concentration (cmc), solubility and definition of the Krafft-point.

The short-range order within the core of a micelle can be characterized as *liquid-like*. It has been shown by NMR-methods (see Section 4.5.2) that for normal micelles close to the micelle surface the orientational order parameter of the CH_2-groups changes only slightly while it decreases drastically towards the terminal CH_n-groups [5]. Consequently, the hydrophobic bulk of a micelle can solubilize quite a lot of organic material. This is one main feature of the technical applicability of surfactant/water systems.

A micelle is by no means a static object. The time scale of micelle formation is about some milliseconds to seconds; a typical residence time of a monomer within a micelle is in the range of micro- to milliseconds [6]. Despite these fairly small time constants, it may take days to weeks until concentrated amphiphile/water mixtures reach their equilibrium state.

Ionic surfactants in an aqueous solvent dissociate completely at small concentrations. Without the shielding action of nearby counterions an aggregation of amphiphilic ions would be impossible, because of the strong repulsive electrostatic forces between neighboring headgroups. Roughly about two-thirds of the counterions can be regarded as bounded to the charged micelle surface. The ion cloud fixes a substantial quantity of hydration water to the aggregate.

It must be mentioned that also non-amphiphilic molecules like certain aromatic dyes can form aggregates in aqueous solvents and in this way exhibit lyotropic liquid crystalline phases, which have been called "chromonics". The aromatic moieties of such molecules are believed to stagger into stacks due to dispersion attraction. Thus, the aggregation driving interaction greatly deviates from that of amphiphilic systems, which manifests itself in a different polymorphism and physical behavior of chromonics [7].

To date, only very few examples of *non-aqueous* lyotropic liquid crystal systems are known; successfully investigated solvents are, e.g., formamide [8], glycerol [9], and ethylene glycol [10]; suitable surfactants are several compounds already well known from aqueous systems. Furthermore, lyotropic polypeptide liquid crystals have been found in organic solvent mixtures [11].

4.3 Liquid Crystalline Structures and Nomenclature

Lyotropic liquid crystals exist in a concentration range between the isotropic (mostly, but not always micellar) solution and the crystalline surfactant. They are mesophases like the thermotropic ones. From a comparison with the latter, it seems straightforward to assume that the structure-building elements, in the case of lyotropics, are anisometric micelles or quasi-micellar aggregates, which take the part of the form-anisotropic molecules of corresponding thermotropic phases. With this assumption an analogous phase sequence could be constructed; the analogy, however, is not very far-reaching, because there are essential differences between the two classes of liquid crystals: e.g., the surfactant aggregates may considerably change their shape and size with temperature and concentration, in contrast to the thermotropic mesogenic molecules.

In the following, a scheme of lyotropic liquid crystalline structures is developed by introducing an increasing amount of order to a surfactant/solvent system. Starting from an isotropic solution of anisometric micelles the first step to achieve a long-range orientational order is to let certain preferred axes of rod-like or disc-like micelles arrange parallel to each other in the time or ensemble average; the resulting phase is called "nematic". The *positions* of the micelles are without any long-range order, as shown in Fig. 4.5a–c. The term "nematic" stems from the Greek word $\nu\eta\mu\alpha$ (thread) which explains itself with the typical appearance of this phase as observed by polarizing microscopy (see Fig. 4.8); it is also the same in the corresponding thermotropic phase (see Chapter 1). Nematic phases of rod-like micelles are denoted as N_C (from "cylindrical" or "calamitic"), while N_D or N_L refers to disc-like micelles. Both nematic phases exhibit long-range orientational order of only one preferred axis. If the two other aggregate axes are not equivalent, it is possible to get these axes ordered, too; then a biaxial nematic ("N_{BX}") results (see Fig. 4.5c). Analogous to the thermotropic cholesteric liquid crystals, also chiral lyotropic phases exist and exhibit a helical structure of the director distribution. Phase chirality can be induced by adding some chiral component to a nematic phase [12].

With increasing order rod-like aggregates can arrange in a two-dimensional *hexagonal* lattice. Then, a long-range positional order occurs in addition to the orientational order of the rods. In a convenient nomenclature the normal hexagonal phase is denoted as H_1 (see Fig. 4.5d). If the rod-like aggregates possess an oval instead of a circular cross-section, a rectangular lattice can arise instead of the hexagonal one.

If, on the other hand, the amphiphile prefers to build up disc-like aggregates, a regular pile of large lamellae, the *lamellar* phase, is to be expected by the addition of positional order to the nematic phase, as shown in Fig. 4.5e. The lamellar lyotropic phases are the counterparts of the thermotropic smectic phases (see Chapter 1). The shorthand symbols for lamellar phases are L_α and L_β. The indices refer to the state of

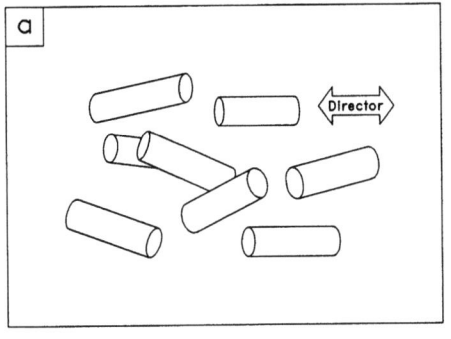

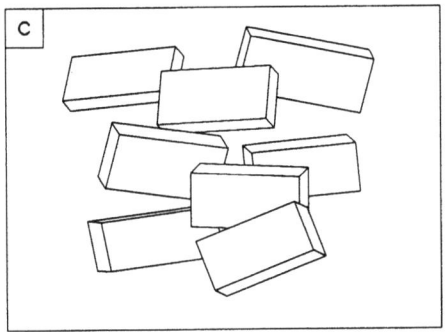
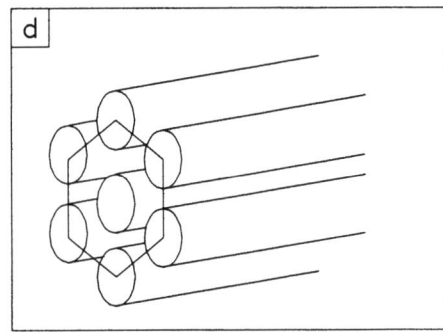
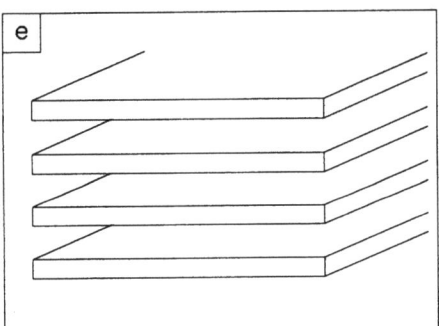
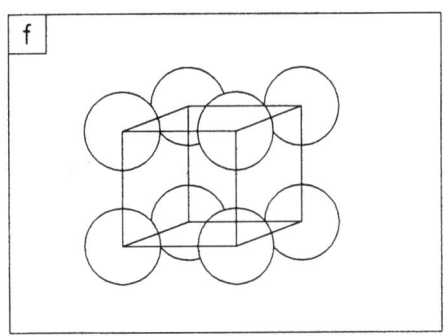
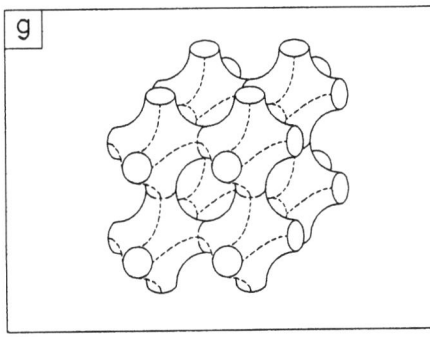

Fig. 4.5. Structure models of lyotropic liquid crystalline phases.
a) Nematic order of prolate aggregates, N_C;
b) Nematic order of oblate aggregates, N_D;
c) Biaxial nematic order of lath-like aggregates, N_{BX};
d) Hexagonal array of long aggregates, H_1, H_2;
e) Lamellar pile of extended discs, L_α, L_β;
f) Cubic arrangement of spherical aggregates, I_1, I_2;
g) Bicontinuous cubic arrangement, V_1, V_2.

the aliphatic chains of the surfactants within the micelles; α stands for liquid-like, disordered chains (containing lots of different conformations), while β means that the chains are in the solid-like all-trans conformation; phases with solid-like alkyl chains are called "gel phases". Occasionally, this term is used in the literature in a sense beyond this definition, to characterize phases with very high viscosity or viscoelasticity, without regard to their structure; this ambiguity should be avoided. Within the lamellae of the gel phases the chains can form a two-dimensional hexagonal lattice, if the space requirement of the headgroups is not too large. Otherwise, a headgroup lattice with disordered tails may result. In addition, the mean orientation of the tails can be perpendicular or tilted with respect to the layer plane. A lamellar phase with tilted crystalline chains is called L'_β. In the sense of the definition given in Section 4.1, also the gel phases (that is: lamellar, with all-trans chains) fulfill the criteria of a liquid crystalline mesophase (long-range orientational order as well as some fluidity); however, it has become convention to distinguish between liquid crystals and gel phases.

The lamellae can consist of bilayers of monomers, thus giving a layer thickness of about or somewhat less than twice the molecular length; also an interdigitation of the molecules in the bilayer may occur.

Cubic lattices built of globular micelles (I_1, I_2) as well as bicontinuous cubic structures (V_1, V_2) have been reported [13]; structure proposals are shown in Fig. 4.5f, g. Regarding their classification, the I-phases own the structure of *plastic crystals* rather than of liquid crystals. If one regards, on the other hand, the surfactant bilayers as the structure-building elements, there is an orientational long-range order (in addition to the positional order) in both phases I and V, of course. In many respects the cubic phases behave like isotropic media; similar cubic structures have been found in thermotropic liquid crystals, occurring as intermediate structures between the cholesteric and the isotropic phase ("blue phases", see Chapter 1).

Unfortunately, the scientific community in the past did not agree to have one uniform nomenclature for the lyotropic phases. A translation between several shorthand symbols is given in [14]. Most frequently used notations are summarized in Table 4.1.

Table 4.1. Shorthand symbols for lyotropic phases.

Isotropic; normal resp. reverse micelles	L_1 resp. L_2
Isotropic; bicontinuous (sponge)	L_3
Hexagonal; normal resp. reverse	H_1 resp. H_2
Nematic; rod-like aggregates	N_C
Nematic; disc-like aggregates	N_D, N_L
Rectangular	R
Lamellar; liquid chains	L_α
Lamellar; all-trans chains	L_β, L'_β
Cubic; spherical aggregates, normal resp. reverse	I_1 resp. I_2
Cubic; bicontinuous, normal resp. reverse	V_1 resp. V_2

4.4 Phase Diagrams

Lyotropic liquid crystals are always two- or more-component systems. The phase diagrams are complex because of the many possible phases as well as the corresponding two- or more-phase regions, according to Gibbs' phase law. Often, the phase boundaries are hardly detectable with standard experimental techniques. Therefore, only a few exact and complete phase diagrams can be found in the literature. However, the obvious validity of the phase law for the systems under discussion shows that lyotropic liquid crystals are real phases in the thermodynamical sense; they are *homogeneous equilibrium systems*. The phase transitions occur at well-defined concentrations and temperatures.

From geometrical arguments (see Section 4.6) a standard phase sequence

isotropic → nematic → hexagonal → lamellar → reverse phases

is to be expected with decreasing water content of the system. Beyond that, in the ideal case the sequence within the reverse phases should be a mirror image of the normal one. Furthermore, several intermediate phases (with cubic and other symmetries) may occur within this sequence [14, 15].

A two-component system will exhibit a phase diagram similar to that of the non-ionic surfactant polyethyleneoxide and water which is shown in Fig. 4.6. Passing through the phase diagram along an isotherm above the solubility temperature, one can cross several liquid crystalline phases. The one-phasic liquid crystal regions in this example are the hexagonal H_1 and the lamellar L_α as well as an intermediate phase (bicontinuous cubic, V_1). The strong polar head group hydration of the polyethyleneoxide results in a broad hexagonal phase region. There is a two-phase region (water + L_1) above the isotropic L_1 phase. The separation of a more concentrated L_1 phase from the solvent phase is caused by a change of the balance between attractive and repulsive inter-micellar forces in favor of the attraction. This can originate from a marked decrease of head group hydration with increasing temperature. On entering this two-phase region (water + L_1) the lyotropic sample gets turbid, due to the phase separation; therefore, the corresponding temperature is called "cloud point". In Fig. 4.6 all of the other two-phase regions have been omitted.

Crossing phase diagrams of lyotropic systems at constant concentration by increasing the temperature, one often finds the phase sequence

crystalline → gel → nematic, or hexagonal, or lamellar, or intermediate → isotropic

In temperature vs. concentration phase diagrams the phase boundaries between different liquid crystalline phases usually are steep, if not vertical, except for the gel to liquid crystalline transition. This shows that the mesophase structure must be due to interactions of low temperature sensitivity. At elevated temperatures the mesophases transform into isotropic phases. This results from the loss of the orientational order by the thermal disordering action which is the analogue to the clearing phenomenon of the thermotropic liquid crystals. For lyotropics the temperature not only relates to

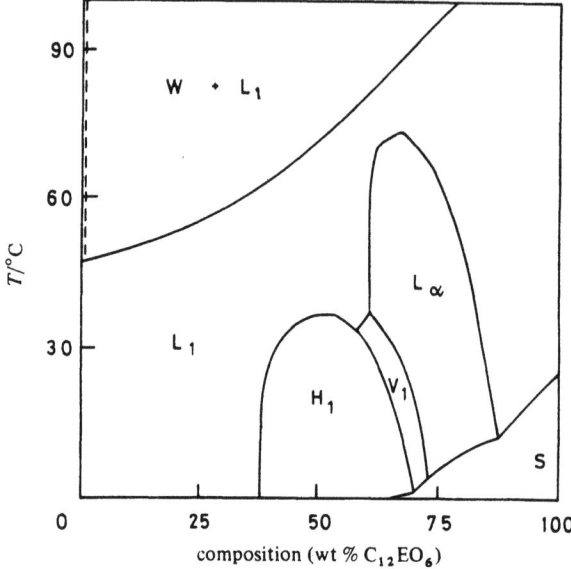

Fig. 4.6. Phase diagram of the polyethyleneoxide $C_{12}H_{25}(OCH_2CH_2)_6OH$/water system; most of the two-phase regions have been omitted. (From [35].)

the thermal motion, but additionally influences the geometry and flexibility of the aggregates via the increased disorder of the long alkyl chains of the amphiphilic molecules, and/or via the temperature-sensitive hydration, etc.

The transitions between different lyotropic phases can be of first or second order. The gel to liquid crystalline transition is of first order and is accompanied by a large enthalpy effect which is easily detected by thermal analysis (order of magnitude 10 kJ/mol [15]). A positional lattice of the surfactant breaks down at this transition(s) due to the "chain melting", or, respectively, the loss of a headgroup lattice (instead of the chain lattice). A lot of different hydrates and crystal modifications may occur below the "Krafft-discontinuity".

Also of the discontinuous type are the nematic to isotropic and the nematic to hexagonal transitions because of symmetry reasons. Nematic to lamellar phase transitions have been found to be of first as well as of second order.

Phase sequences of course are not always as clear-cut as in Fig. 4.6. Fig. 4.7 is a three-component diagram which has been observed at constant temperature. Compared to the minimum-component system (solvent + surfactant) a "co-surfactant" has been added as the third component; this is frequently used to adjust the aggregate shape, and further properties, as will be explained in Section 4.6. In this diagram all experimentally found two-phase regions are included. Additionally different reverse phases at low water content are shown. The dominant lamellar phase extends from about 85 to 25 wt% surfactant (incl. co-surfactant), i.e., from 15 to

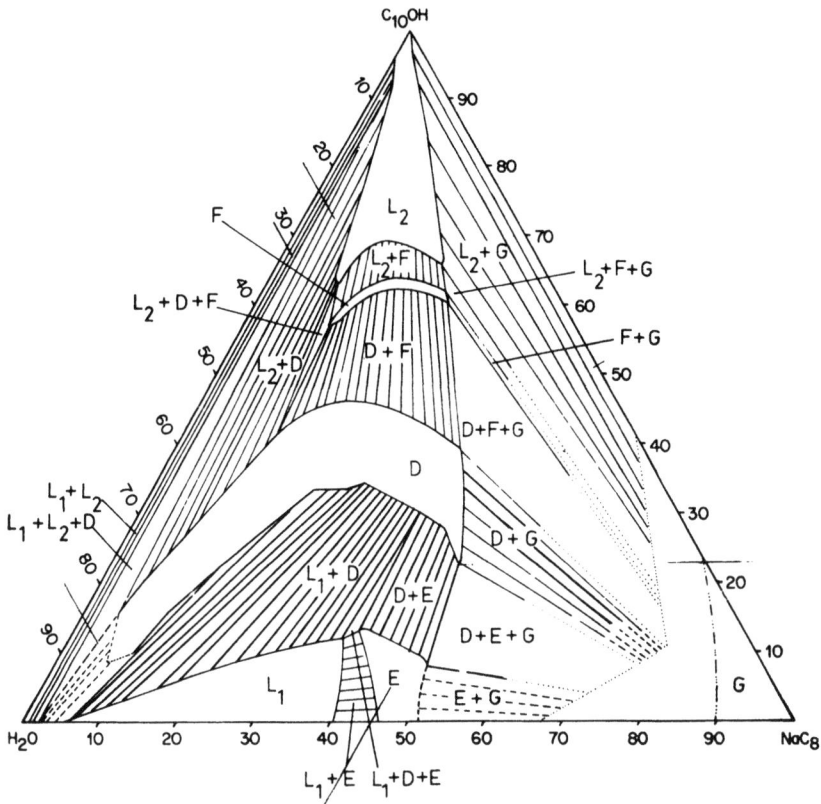

Fig. 4.7. Complete phase diagram of sodium octanoate/water/decanol ($NaC_8/H_2O/C_{10}OH$). $D \equiv L_\alpha$, $E \equiv H_1$, $F \equiv H_2$, $G \equiv$ hydrated crystalline. (From [36]; see also Table 4.1)

75 wt% water content. This indicates that the lamellar phase can swell considerably by the uptake of water. In principal, the other phases can swell, too. Especially for the lamellar phase this swelling phenomenon has been observed on a very large scale, in extreme cases up to surfactant concentrations below 1 wt% and inter-lamellae distances of more than 500 nm. It seems surprising that even at this enormous amount of intercalated water lamellar phases can be stable. Theoretical aspects of the extreme swelling are discussed in Section 4.6.

To date, there are still only very few examples of nematic lyotropic phases. Mostly the nematic phase regions are narrow; concentration intervals of about 5 to 10 wt% and temperature ranges of 20 K have been found. Their stability seems to be small [16].

155

4.5 Physical Properties and Investigation Methods

Due to their orientational long-range order, liquid crystals usually are uniaxial or biaxial *anisotropic* media. Among the many anisotropic properties are the dielectric and the diamagnetic susceptibility, the viscosity, and the light refraction. The optical anisotropy of lyotropic phases is small, compared to that of thermotropics; it is of the order of 10^{-3}. The magnitude as well as the sign of the birefringence are determined by some contribution of the *molecular* and by an additional (small) one of the *aggregate* anisotropy ("form anisotropy"). The signs of the two contributions are opposite in most cases, due to the fact that the long axes of the amphiphilic molecules (and mostly therewith of the electronic polarizabilities) are, on the average, perpendicular to the long axes of the aggregates. In most cases prolate, respectively, oblate structures possess negative, respectively, positive birefringence, caused by the specific average orientation of the electronic polarizability ellipsoid of the molecule. Exceptions from this rule may occur due to aromatic moieties of the surfactant compound. A biaxial nematic phase can be formed by biaxial micelles [17].

In reality, the director orientation of a liquid crystal is never completely uniform (see Figs. 4.5, 4.8–4.10). Due to a fixed anchoring of surfactant monomers and/or aggregates at the container surfaces, the local director orientation varies within the surface as well as the bulk of a sample. Without external fields the orientation is determined by the elasticity of the phase. This often only metastable sample topology is called "texture". In order to minimize the elastic deformation energy, the director prefers to vary only smoothly with position; however also well-defined abrupt orientation changes occur, the so-called "disclinations" [18]. Because of the birefringence coupled to the director, the textures of lyotropics, like that of thermotropics, can be observed by means of crossed polarizers. Areas of equal optical path difference between ordinary and extra-ordinary beam appear with the same brightness and/or color. They are called "Schlieren" (see Fig. 4.8). A first qualitative classification of a sample can be performed with the naked eye with a small sample in a glass tube. Thus, an anisotropic phase can be simply distinguished from an isotropic one; occasionally, shear-induced birefringence can be detected. From a rough estimate of the viscosity and the birefringence of a sample, the observer gets a (very preliminary) hint about the phase type: readily flowing samples with colored "Schlieren" are presumably nematic; enhanced viscosity and apparently colorless birefringence (no Schlieren) points to lamellar, and scarcely flowing samples with apparently colorless birefringence might be hexagonal.

4.5.1 Polarizing Microscopy

Some more details of a sample can be recognized using a polarizing microscope. Indeed, every liquid crystalline phase can exhibit certain typical textures, as is explained in more detail in Chapter 1 for the thermotropics; but, unfortunately, the observed appearances are not always clear-cut in identifying a phase without doubt. Beyond that, isotropic phases of statistically oriented anisometric micelles may be

Fig. 4.8. "Schlieren" texture of a lyotropic nematic phase N_L.

forced to exhibit shear-induced orientational order. In the case of elevated viscosity, or viscoelasticity, such a system cannot relax to the equilibrium isotropy and, thus, a liquid crystalline structure is simulated. A lot of experimental experience is necessary and, in any case, additional methods have to be used for an unequivocal determination of a phase type.

In the following only the clearest and typical textures to be seen in a polarizing microscope are briefly discussed. A detailed description is given in [19]. Nematic phases – thermotropic as well as lyotropic ones – often exhibit a Schlieren texture immediately after preparation of a sample, due to a non-uniform flow alignment of the director field. The occurrence of a Schlieren texture is a distinct hint to the nematic phase. An example of this is shown in Fig. 4.8. Without flow, nematics N_D usually tend to a "homeotropic" orientation: the director and the optical axis of the sample then turn into a perpendicular alignment with respect to the slides so that the sample appears black with orthoscopic observation, but may be well distinguished from an isotopic phase by conoscopic observation. Rod-like nematics, however, either remain in the Schlieren texture or transform into a homogeneous planar or

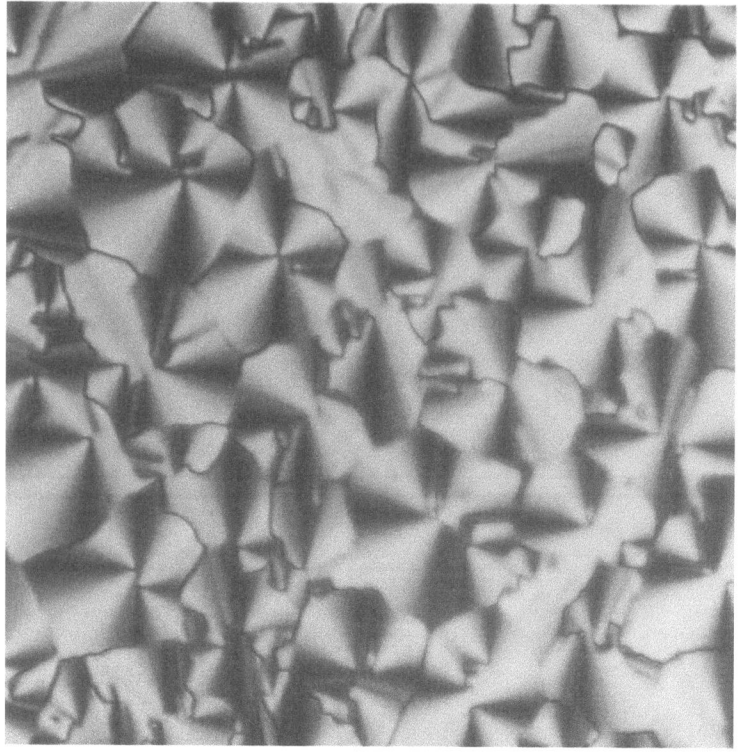

Fig. 4.9. Fan texture of a lyotropic hexagonal phase H_1.

tilted orientation by action of the boundary solid surface; in most cases, they do not align homeotropically.

The hexagonal phase is best identified by a "fan" texture, which is a consequence of the prevention of a uniform alignment due to the rapid growth of several nuclei at the phase transition or by turbulent flow. The curved alignment then follows "focal conic" surfaces to reach an arrangement of locally minimal strain. The typical texture is shown in Fig. 4.9, see also [18].

A lamellar phase strongly tends to a homeotropic alignment, as the N_D nematic does. Fortunately, these two phases can be well distinguished by the observation of "oily streaks", which are unique for lamellar phases. Oily streaks are bundles of disclinations within the layer structure (see Fig. 4.10; see also [18]).

4.5.2 Nuclear Magnetic Resonance

Nuclear spins are local probes. Their energy levels are sensitive to their environment through dipolar and quadrupolar interactions. Today, NMR is a very well developed

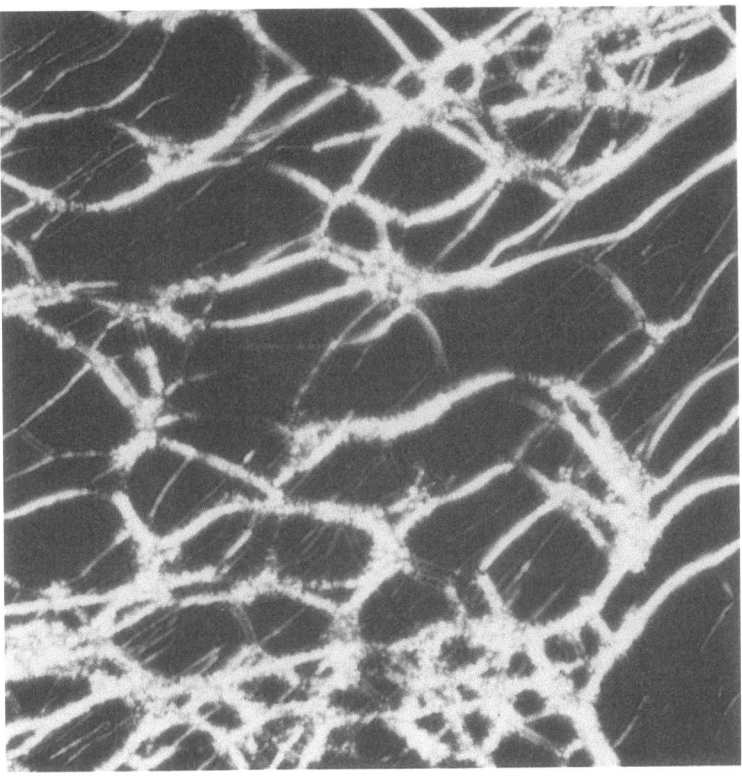

Fig. 4.10. Homeotropic texture of a lyotropic lamellar phase L_α with "oily streaks".

technique which allows to study almost every nucleus of the periodic table. In addition, several parameters of a probe material can be measured; among these are the chemical shift and relaxation rates. Thus, many aspects of lyotropic liquid crystals can be investigated by NMR. Examples are: phase diagrams, counterion binding, solubilization, structure of isotropic and anisotropic phases (chain packing, micelle size and shape), and molecular dynamics [20].

Due to their diamagnetic anisotropy, lyotropic liquid crystals can be oriented uniformly by magnetic fields, if an elevated viscosity does not prevent this. Both signs of diamagnetic anisotropy occur. Typical values are about $10^{-11}\,m^3\,kg^{-1}$ which, again, is only a hundredth of the value of corresponding thermotropic phases (see Chapter 2.5). In practice, only nematic phases are alignable by accessible field strengths. Yet an almost uniform alignment of lamellar and hexagonal phases can be reached by the action of flow, and by surface effects at the container walls. Therefore, the molecular motion is anisotropic in suitably large domains; hence, not all dipolar

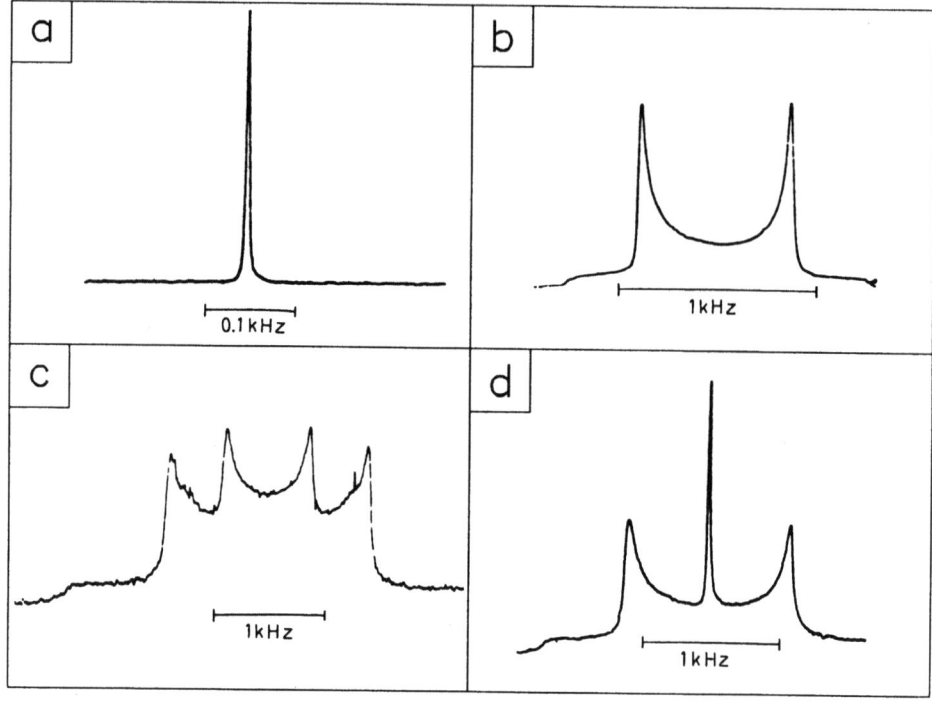

Fig. 4.11. Examples of deuterium NMR spectra of a lyotropic liquid crystalline system. (From [20].) a) Isotropic phase; b) Anisotropic phase; c) Mixture of two anisotropic phases, presumably hexagonal and lamellar; d) Mixture of isotropic and anisotropic phase.

and quadrupolar interactions are averaged to zero. As an example, some deuterium spectra of different lyotropic phases are given in Fig. 4.11. There is no quadrupole splitting in isotropic phases: a singlet is observed, as in Fig. 4.11a. Figure 4.11b is symptomatic for an anisotropic phase; the concrete phase type cannot be concluded from this spectrum alone, however. Figure 4.11c results from a two-phase system, both phases being anisotropic. This finding points to the presence of a hexagonal/lamellar equilibrium, because the lamellar phase can be expected to exhibit a quadrupolar splitting of about twice the hexagonal one. The peaks in Fig. 4.11d originate from an isotropic/anisotropic two-phase sample. The clear detection of both phases in a two-phase region is an advantage of the NMR-technique, especially if one starts searching for some distinct phase in an unknown system.

4.5.3 Small-Angle Scattering of X-rays (SAXS) and of Neutrons (SANS)

Final clearness about the structure of a phase often can be achieved only by the interpretation of interference patterns of waves with suitable wavelengths, scattered by the sample. In the periodicity range occurring in lyotropic liquid crystals, SANS with thermal neutrons as well as SAXS are very well suited methods. Thermal (≈ 300 K) neutrons have a velocity of ≈ 2200 m s^{-1}, corresponding to a de Broglie wavelength of 0.2 nm. This is also in the range of x-ray wavelengths. Detectable particle sizes are 5 to 500 nm for SANS, while one is restricted to a maximum of about 30 nm for usual SAXS-cameras; however, this is, in many cases, sufficient to identify liquid crystalline phases.

Neutron scattering is due to nuclear and magnetic interaction, while x-rays are sensitive to electron density distributions. High electron densities produce an intensive scattering. Many materials are transparent for neutrons; the cross-section effective for a scattering process varies strongly from one element or isotope to another (e.g., from ^{1}H to ^{2}H), due to complex dependencies. This phenomenon is utilized for contrast variation (and contrast matching), which allows to see, for example, only the core of a micelle by adjusting the proportion H_2O/D_2O in a liquid crystalline solution [21].

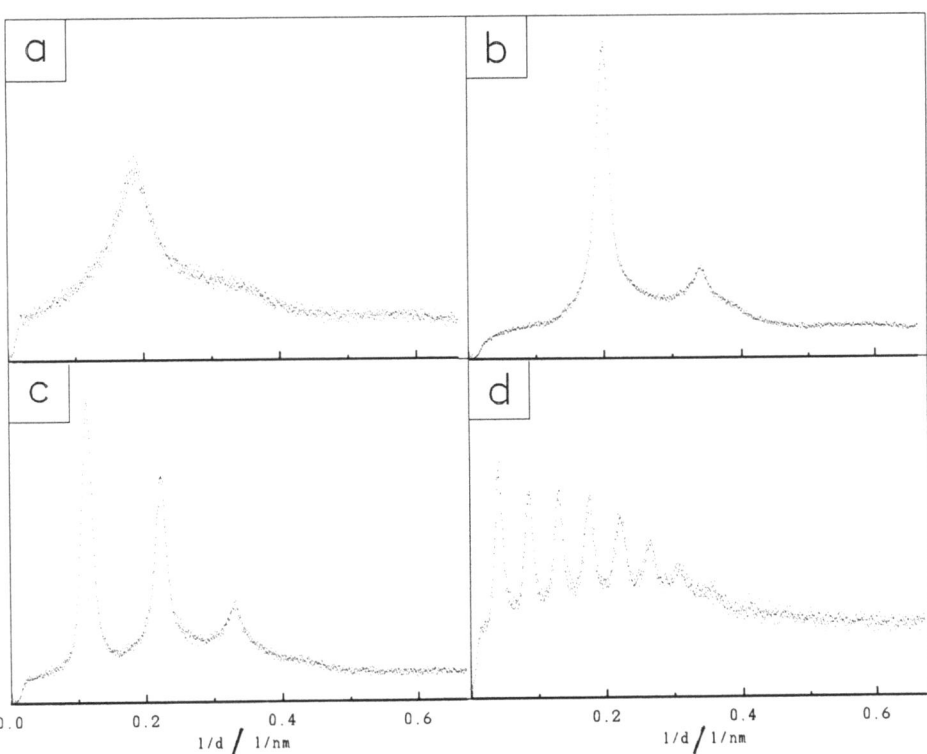

Fig. 4.12. Typical x-ray scattering results of lyotropic liquid crystals.
a) Nematic phase N_C; b) Hexagonal phase H_1; c) Lamellar phase L_α; d) Lamellar phase L_β.

Some x-ray scattering results are shown in Fig. 4.12. Nematic phases show only one broad scattering peak; reflexes at the distinct proportions $1:\frac{1}{2}:\frac{1}{3}:\frac{1}{4}\ldots$ of the scattering vectors are due to a lamellar structure, and a hexagonal lattice of rods is concluded from peaks at $1:\frac{1}{\sqrt{3}}:\frac{1}{\sqrt{4}}:\frac{1}{\sqrt{7}}\ldots$. Further parameters which can be deduced from the spectra of a liquid crystalline sample are the mean lateral area of the chains and headgroups, the diameter of rod-like micelles, and the bilayer thickness of lamellar phases. The two latter values have been found to always be 10 to 50% less than twice the maximum molecular length.

For the intermediate phases a lot of cubic structures, belonging to different space groups, have been recognized by x-ray scattering [13].

4.6 Theoretical Aspects

The problem to explain and predict phase diagrams of lyotropic liquid crystals has not yet been perfectly solved. Two- or more-component systems are not easily handled, experimentally or theoretically. But there are several concepts which supply a lot of basic understanding about the features of lyotropics.

4.6.1 Interactions

In the following the various types of inter- and intra-micellar interactions are itemized; some of them contribute even at long distances, others are only very short-reaching. The forces may act *within*, as well as *between* aggregates [25].

Van der Waals dispersion forces

Dispersion forces may, at the first stage, be understood as an interaction of fluctuating dipoles, occurring also between *nonpolar* atoms. The van der Waals attraction potential between two atoms at distance d in vacuum has the form

$$W = -\text{const.}/d^6.$$

Poly-atomic particles obey a more complicated power law. In most cases the forces are attractive, but they may be repulsive at certain dielectric properties of the medium between the interacting particles. The forces generally are anisotropic, can be relatively long ranged (to about 10 nm), and depend mainly on the interacting material; especially, they are not very sensitive to salt concentration and to pH changes of the solvent. For example, for two planar surfaces at distance d the free energy W per unit area is

$$W = -H/12\pi d^2,$$

where H is the Hamaker constant, containing the polarizability of the interacting particles. This power law is valid only for short distances; for farther apart lamellae W decays with d^{-4}.

Electrostatic forces

Electric charges, present in the polar or ionic headgroups of amphiphiles, interact via the Coulomb force. To allow for aggregation, sufficient screening through the presence of opposite charges is essential. Aggregates of ionic surfactants are surrounded by diffuse layers of counterions. There is a more or less strong repulsion between the electrostatic double layers of neighboring aggregates; the effective range of influence depends on the diffusity of the layer and is characterized by the Debye length. The diffusity can be adjusted by the electrolyte concentration of the surfactant/solvent system. By addition of salt the Debye length may be reduced to less than 1 nm. As an example, the electrostatic interaction energy for two planar, equally charged surfaces is

$$W = \text{const.}\ \sigma^2 \exp(-\kappa d)/\kappa\varepsilon,$$

where σ is the surface charge density, κ^{-1} the Debye length, and ε the dielectric constant of the medium between the surfaces at distance d.

Solvation forces

In aqueous systems hydration of the surfactant hydrophilic headgroup occurs. Thereby water molecules in the vicinity of the headgroups are bounded to them rather strongly due to ion/dipole or dipole/dipole interaction. The nearest water molecules build up the primary hydration shell, and further shells may follow. For a close approach of two hydrated aggregates, parts of the hydration shells have to be removed which may cost several hundred kJ/mol. The decay length of this kind of interaction is only about 0.3 nm.

Hydrophobic attraction

Hydrophobic surfaces in water are surrounded by shells of ordered water molecules. A close approach of two such surfaces leads to an overlap of these shells, which therewith reduces the entropically unfavored ordered zone. The "hydrophobic effect" thus results in an attractive interaction between hydrophobic molecules or surfaces in water (which was detailed in Section 4.2).

Steric interactions

Besides the short-ranged steric repulsion between single surfactant molecules as well as between aggregates also a long-ranged steric repulsion between lamellar aggregates has been derived theoretically [26]. This can arise by undulations of the lamellae; it has been predicted to be even farther reaching than van der Waals attraction and electrostatic repulsion. Lamellar phases have proved to be stable even at very low surfactant concentrations down to ≈ 1 wt% [27]. Then the lamellae are about 100 nm apart from each other. Possibly, undulation forces are suitably far reaching to explain this extreme swelling behavior.

4.6.2 The R-theory

After Winsor, the "R-theory" [22] models phase types through the consideration of aggregate shapes. The reference phase type is the lamellar one which consists of alternate, flat, two-dimensionally extended regions of hydrophobic ("oil") and hydrophilic ("water") character. All of the other types of lyotropic liquid crystals can be classified by belonging either to the "normal" or to the "reverse" type (see Section 4.2.2); in normal phases the amphiphilic layers are convex towards the oil region, in reverse phases they exhibit the opposite curvature. Within the theory a parameter R is defined as the tendency C_O of an amphiphilic layer to become convex towards the oil region of a solution, relative to the tendency C_W to become convex towards the water region:

$$R = \frac{C_O}{C_W}$$

Clearly, $R = 1$ results for a plane lamellar phase. The normal liquid crystalline phases are described by $R \leq 1$, while $R \geq 1$ is attributed to the reverse phases. Obviously the R-parameter depends on the strength of the hydrophilic and hydrophobic character of the surfactant. Its respective value is influenced by the interactions between the surfactant and the hydrophilic as well as the hydrophobic medium. The interaction can be quantified by the corresponding interfacial energies. In that it is related to the hydrophile/lipophile balance ("HLB-parameter") of an amphiphile. The dimensionless HLB-parameter of a given surfactant is a measure for the free energy to put the amphiphilic material into an oil/water interface; this parameter was introduced to characterize the emulsification power of a surfactant [23, 24]. It is empirically accessible and ranges between 1 and 20. For example, a HLB-value of about 10 is assigned to an amphiphile that tends to form flat bilayers.

4.6.3 Packing Considerations

A more detailed consideration of the effect of the aggregate shape was given by Israelachvili [25]. His results are summarized in Fig. 4.13 and will be briefly explained in the following. A relatively large headgroup space requirement results in a strong curvature of the aggregate surface; normal, spherical micelles are preferred in that case. On the other hand, spherical micelles bring about more intensive contact of water and hydrocarbon chains, which reduces the entropy of the water (see Section 4.2).

Lipid	Critical packing parameter $v/a_0 l_c$	Critical packing shape	Structures formed
Single-chained lipids (surfactants) with large head-group areas: SDS in low salt	<1/3	Cone 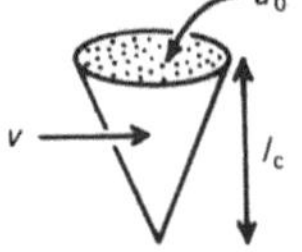	Spherical micelles 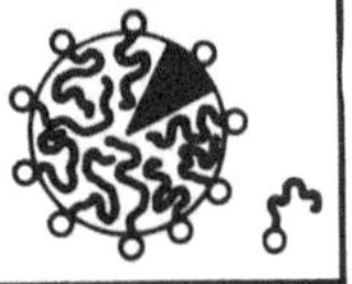
Single-chained lipids with small head-group areas: SDS and CTAB in high salt, nonionic lipids	1/3–1/2	Truncated cone 	Cylindrical micelles 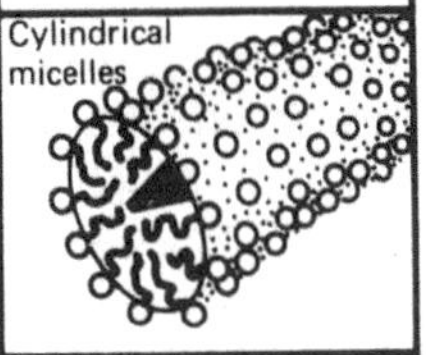
Double-chained lipids with large head-group areas, fluid chains: Phosphatidyl choline (lecithin), phosphatidyl serine, phosphatidyl glycerol, phosphatidyl inositol, phosphatidic acid, sphingomyelin, DGDG[a], dihexadecyl phosphate, dialkyl dimethyl ammonium salts	1/2–1	Truncated cone 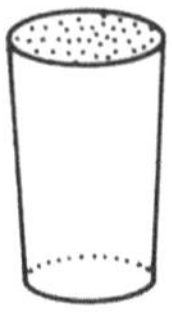	Flexible bilayers, vesicles 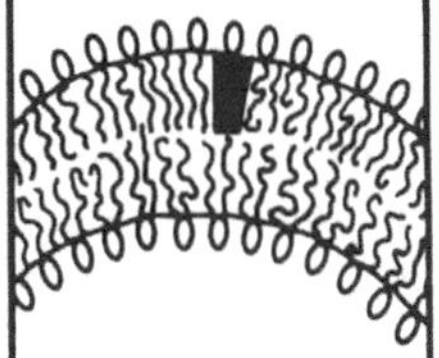
Double-chained lipids with small head-group areas, anionic lipids in high salt, saturated frozen chains: phosphatidyl ethanolamine, phosphatidyl serine + Ca^{2+}	~1	Cylinder 	Planar bilayers 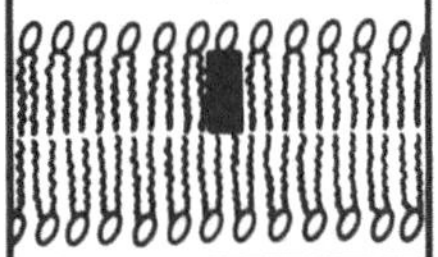
Double-chained lipids with small head-group areas, nonionic lipids, poly (cis) unsaturated chains, high T: unsat. phosphatidyl ethanolamine, cardiolipin + Ca^{2+} phosphatidic acid + Ca^{2+} cholesterol, MGDG[b]	>1	Inverted truncated cone or wedge 	Inverted micelles 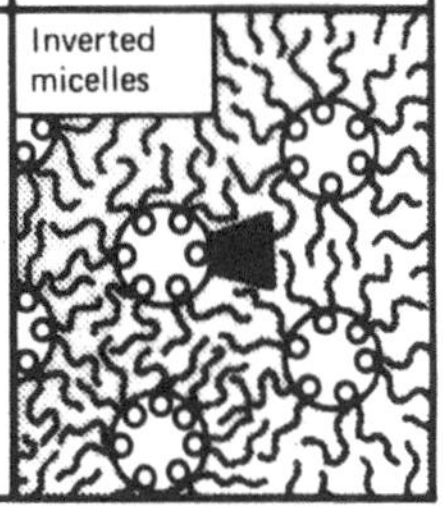

[a] DGDG, digalactosyl diglyceride, diglucosyl diglyceride; [b] MGDG, monogalactosyl diglyceride, monoglucosyl diglyceride.

Fig. 4.13. The packing of amphiphilic molecules in dependence of the "packing parameter" $(v/l_c) : a_o$. (From [25].)

The hydrophobic tail of an amphiphilic molecule is geometrically characterized by its volume v and a length l_c, which is about the all-trans length of the hydrocarbon chain. The headgroup of the molecule requires a certain area a_0 within a hydrophilic/hydrophobic interface. The value of this area a_0 appears as a compromise between the attractive inter-molecular interactions driving the aggregation, and the steric and electric headgroup repulsion on the other side. The point is now the definition of a "packing parameter", which is the ratio of the apparent cross-section of the hydrophobic tail (v/l_c) to the headgroup area a_0. As a rule of thumb, every aggregate shape corresponds to a certain range of $(v/l_c):a_0$, which can be calculated from simple geometrical considerations, just by regarding the volume/surface ratios for spheres, cylinders and lamellae. Normal, spherical micelles will form for $(v/l_c):a_0 \leq \frac{1}{3}$, rod-like aggregates for $\frac{1}{3} \leq (v/l_c):a_0 \leq \frac{1}{2}$, and bilayers for $\frac{1}{2} \leq (v/l_c):a_0 \leq 1$. Values of the packing parameter larger than one will lead to reverse micelles. The packing parameter is temperature-dependent (e.g., by the temperature dependence of the headgroup hydration); for ionic surfactants it is, of course, also influenced by the electrolyte content of the solvent; thus, the shape and size of an aggregate are adjustable to a certain amount.

The shape of a micelle can also be changed by the addition of further components to the system. A "co-surfactant" is an amphiphilic material which mixes with the surfactant in the micelle. Mostly, it is a molecule with a relatively small polar head group, but with an alkyl chain similar to that of the surfactant; alcohols are frequently used. Such co-surfactants reduce the mean headgroup area. Consequently, the preferred surface curvature changes; a spherical micelle may transform into a cylindrical one, a rod turns into a lamella. With ionic surfactants a similar effect is reached by the addition of salt which reduces the Coulomb repulsion of the headgroups.

On increasing the concentration of the surfactant the shape of the micelles must change, too, due to the repulsive interaction between the aggregates. At a fixed concentration, spheres have the smallest inter-micellar distance, compared to rods or discs. Hence, to minimize its free energy, a lyotropic system can change from globular to rod-like or disc-like micelles.

As a further way to maximize their distances, the micelles can arrange into positional and orientational lattices. Transitions to cubic phases of globular micelles, to hexagonal phases of long rods, and to lamellar phases of large discs, are expected.

4.6.4 Other Theories

The interplay of van der Waals attraction and electric double-layer repulsion of charged surfaces has been considered by Derjaguin, Landau, Verwey, and Overbeek in their "DLVO"-theory, which deals with the stability of colloidal suspensions [28]. The potential energy versus distance function is shown in Fig. 4.14. It does not take into account the short-ranged forces. This theory is insufficient for many aqueous systems, and especially for lyotropic liquid crystals.

Other early theories only regard interactions between *rigid* particles, yet, they are able to describe an isotropic/nematic phase transition of rigid macromolecules and of

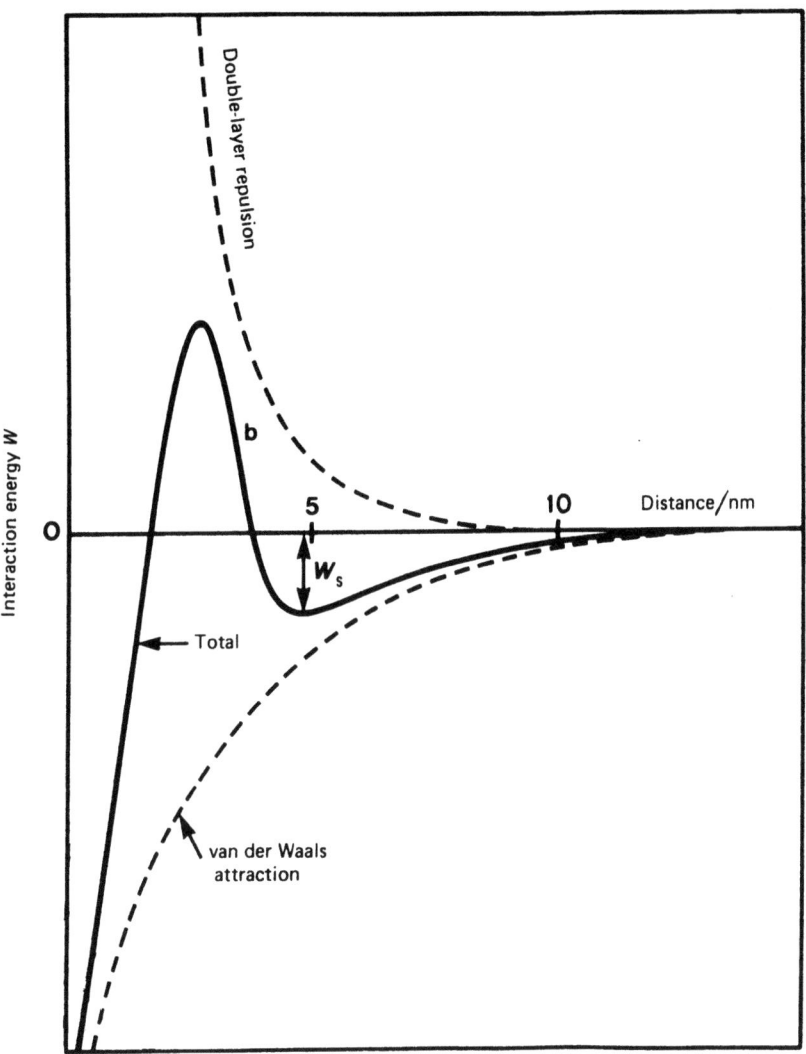

Fig. 4.14. Schematic potential functions of DLVO interaction. (After [25].)

"conventional" colloids, which, in contrast to micellar systems, consist of strongly bounded molecules [29]. Therefore, the used potential functions are not rigorously applicable to real lyotropic systems because the amphiphilic aggregates usually do not behave like solid rods or discs. Rather, they can be very flexible and, in addition, they can change their size. Not only inter- but also intra-aggregate interactions have to be taken into account. Furthermore, there is an interplay between both of them. The consequence of this interplay is a *coupling* between the orientational order parameter $S = \frac{1}{2}\langle 3\cos^2\theta - 1\rangle$ and the size and shape of the aggregates (θ = angle between director and preferred aggregate axis, see also Chapter 1). It has been shown theoretically that an anisotropic excluded volume interaction significantly enhances the length-to-width ratio of rod-like aggregates, which then again increases the anisotopy of this interaction [30] (excluded volume interaction see also Chapter 1). In other words, the structure building elements reorganize themselves internally through a transition to an orientationally ordered phase. If the coupling between the orientational order and the dimensions of the aggregates is strong enough, according to the theory the elongation of the rods may diverge; a nematic phase may thus transform into a hexagonal one. However, some actual experiments show that the formation of very elongated rods is not a necessary precondition for the hexagonal order.

A concept very different from the above-mentioned was proposed by Charvolin [31]; it emphasizes the essential role of the amphiphilic film as the structure-building element. The theory avoids a calculation of the free energy from several intra- and inter-aggregate forces which mostly are not known in detail. Instead it regards the components of the overall interaction normal respectively parallel to the films; the normal ones act to maintain a constant distance between the films, the parallel ones intend a distinct film curvature. The model basically takes into account the *frustration* arising from these interactions which are necessarily conflicting, if the films are arranged in periodic configurations. Optimization leads to distinct arrangements of the films resp. of the disclinations which have to be introduced, e.g., by a Volterra process [32]. The resulting structures indeed are closely related to the experimentally established ones.

4.7 Applications

Compared to thermotropic liquid crystals the lyotropic ones to date have found only a few applications. Thermotropic liquid crystalline phases serve as the active medium in display devices (see Chapter 6). This capability of the mesophases results from the unique combination of liquid-like and crystal-like properties. The well-developed display technology exclusively uses electric field effects which have the advantage of extremely low power consumption and years of faultless operation. But to date, the application of lyotropic liquid crystals in display devices fails due to the fact that the common solvent is water with its relatively large electrical conductivity. Moreover, only a slow switching can be expected. However, the use of suitably modified lyotropic systems would be greatly appreciated for economical and ecological reasons. Some effort seems to be necessary to discover well suited non-aqueous lyotropic systems which can compete with the sophisticated thermotropic materials.

An actual application of lyotropic mesophases is due to their relationship to biological membranes which consist of lipid bilayers [33]. Essential properties of cell membranes originate from their liquid crystalline behavior. In this respect dilute lamellar phases or dispersions of lamellae in water are valuable model systems for the biological research.

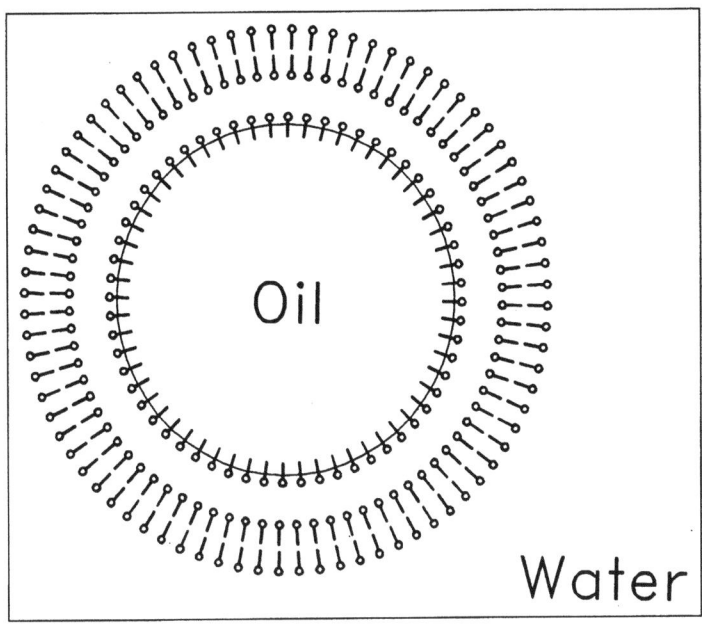

Fig. 4.15. Lamellar surfactant multilayer stabilizing an oil droplet in a hydrophilic medium.

Surfactant bilayers must not be plane in any case. They can grow crooked and thus may form a closed shell, called a "vesicle". These *micro-vessels* have the capability to encapsulate suitable materials present in the solution. The core of a normal vesicle is an *aqueous* phase; but also the stabilization of an *oil* droplet by a surrounding lamellar multilayer can be achieved (see Fig. 4.15). Such micro-vessels are capable of carrying hydrophilic as well as lipophilic compounds through aqueous media and protecting them from outside influences. These features are increasingly utilized for pharmaceutical and cosmetic purposes, to realize a well controlled activity of drugs.

A further application is the control of chemical reactions occurring within a liquid crystalline host phase. As an example, the influence on the reaction kinetics has been shown by changing the size and/or the shape of the mesogenic aggregates via surfactant or electrolyte concentration and of temperature [34].

Also, the capability to extract and isolate hydrophobic compounds from aqueous systems should be mentioned. This procedure has been proved to work with reverse micelles and with lamellar phases. The small capacity of the lyotropic phases restricts this application to especially valuable compounds.

References

1. Forbes RJ (1965) Studies in Ancient Technology. Brill, Leiden, Netherlands
2. Tanford C (1980) The Hydrophobic Effect. Wiley-Interscience, New-York

3. Quist PO, Halle B, Furó I (1991) Micelle Size and Order in Lyotropic Nematic Phases from Nuclear Spin Relaxation. J Chem Phys 96: 3875–3891
4. Ekwall P (1972) Aggregation in Surfactant Systems. Surface Chemistry and Colloids, Physical Chemistry Series 1, 7: 98–145
5. Gruen DWR (1985) A Model for the Chains in Amphiphilic Aggregates. J Phys Chem 89: 146–163
6. Bauernschmitt D, Hoffmann H, Platz G (1981) Zur Dynamik des Aggregationsverhaltens zweikettiger ionogener Tenside. Ber Bunsenges Phys Chem 85: 203–210
7. Attwood TK, Lydon JE, Hall C, Tiddy GJT (1990) The Distinction between Chromonic and Amphiphilic Lyotropic Mesophases. Liqu Cryst 7: 657–668
8. Jonströmer M, Sjöberg M, Wärnheim T (1990) Aggregation and Solvent Interaction in Nonionic Surfactant Systems with Formamide. J Phys Chem 94: 7549–7555
9. Auvray X, Petipas C, Anthore R, Rico I, Lattes A (1989) X-ray Diffraction Study of Mesophases of Cetyltrimethylammonium Bromide in Water, Formamide, and Glycerol. J Phys Chem 93: 7458–7464
10. Friberg SE, Ward AJI, Larsen DW (1987) Dynamic Structure of a Nonaqueous Lamellar Liquid Crystal: Comparison with the Aqueous Case. Langmuir 3: 735
11. Samulski TV, Samulski ET (1976) Van der Waals-Lifshitz Forces in Lyotropic Polypeptide Liquid Crystals. J Chem Phys 67: 824–830
12. Melnik G, Saupe A (1987) Microscopic Textures of Micellar Cholesteric Liquid Crystals. Mol Cryst Liq Cryst 145: 95–110
13. Luzzati V, Mustacchi H, Skoulios AE, Husson F (1960) La structure des colloides d'association. I. Les phases liquide-cristallines des systemes amphiphile-eau. Acta Cryst 14: 660
14. Tiddy GJT (1980) Surfactant-Water Liquid Crystal Phases. Physics Reports 57: 1–46
15. Kékicheff P, Grabielle-Madelmont C, Ollivon M (1989) Phase Diagram of Sodium Dodecyl Sulfate-Water System. J Colloid Interf Sci 131: 112–132
16. Tezak D, Hertel G, Hoffmann H (1991) Phase Behaviour of N-alkyl and Bi-alkylbenzene Sulphonates. Nematic Lyotropics from Double Chain Surfactants. Liqu Cryst 10: 15–27
17. Yu LJ, Saupe A (1980) Observation of a Biaxial Nematic Phase in Potassium Laurate-1-Decanol-Water Mixtures. Phys Rev Lett 45: 1000–1003
18. Demus D, Richter L (1978) Textures of Liquid Crystals. Verlag Chemie, Weinheim
19. Rosevear FB (1954) The Microscopy of the Liquid Crystalline Neat and Middle Phases of Soaps and Synthetic Detergents. J Am Oil Chem Soc 31: 628–638
20. Lindman B, Stilbs P (1985) Nuclear Magnetic Resonance of Surfactant Systems, Physics of Amphiphiles: Micelles, Vesicles and Microemulsions. Elsevier, Amsterdam
21. Stuhrmann HB (1979) Neutronenstreuung an Biopolymeren. Chem in unserer Zeit 13: 11–22
22. Winsor PA (1968) Binary and Multicomponent Solutions of Amphiphilic Compounds. Solubilization and the Formation, Structure, and Theoretical Significance of Liquid Crystalline Solutions. Chem Rev 68: 1–40
23. Gruen WR, Lacey EHB de (1984) The Packing of Amphiphile Chains in Micelles and Bilayers. Surfactants in Solution 1: 279–306
24. Becher P (1984) HLB – a Survey. Surfactants in Solution 3: 1925–1946
25. Israelachvili JN (1985) Intermolecular and Surface Forces. Academic Press, London
26. Helfrich W (1977) Steric Interaction of Fluid Membranes in Multilayer Systems. Z Naturforsch 33: 305–315
27. Appell J, Bassereau P, Marignan J, Porte G (1989) Extreme Swelling of a Lyotropic Lamellar Liquid Crystal. Colloid Polym Sci 267: 600–606
28. Verwey EJW, Overbeek JTG (1948) Theory of Stability of Lyophobic Colloids. Elsevier, Amsterdam

29. Flory PJ, Ronca G (1979) Theory of Systems of Rodlike Particles. Mol Cryst Liqu Cryst 54 : 289–330
30. Gelbart WM, Ben-Shaul A, Masters A, McMullen WE (1985) Effects of Interaggregate Forces on Micellar Size in Isotropic and Aligned Phases, Physics of Amphiphiles: Micelles, Vesicles and Microemulsions. Elsevier, Amsterdam
31. Charvolin J (1989) Lyotropic Liquid Crystals, Structures and Phase Transitions. NATO ASI Ser. B 211 : 95–111
32. De Gennes PG (1974) The Physics of Liquid Crystals. Clarendon Press, Oxford, p 262
33. E.g., Bolis L, Pethica BA (editors) (1968) Membrane Models and the Formation of Biological Membranes. North Holland, Amsterdam
34. Ramesh V, Labes MM (1987) Bimolecular Reactions in Nematic Lyotropics: A Study of Catalized Ester Hydrolysis in N_c, N_l, and I Phases. Mol Cryst Liqu Cryst 144 : 257–261
35. Mitchell DJ, Tiddy GJT, Waring L, Bostock T, McDonald MP (1983) Phase Behaviour of Polyethylene Surfactants with Water. J Chem Soc, Faraday Trans 79 : 975
36. Friman R, Danielsson I, Stenius P (1981) Lamellar Mesophase with High Contents of Water: X-ray Investigations of the Sodium Octanoate-Decanol-Water System. J Colloid Interf Sci 86 : 501–514

5 Application of Liquid Crystals in Spectroscopy

L. Pohl

Introduction

The spectroscopic properties which are required for the characterization of organic molecules and, in particular, for the structure elucidation of unknown molecules are, as a rule, independent of direction. Measurements, e.g., UV, IR, NMR, ESR, OR, ORD and CD are, hence, normally carried out in isotropic solvents in which a preferred molecular orientation cannot take place.

However, a number of anisotropic molecular properties can only be observed with oriented molecules, e.g., the polarization of an electronic transition in the UV, a vibrational transition in the IR or the magnitude of a direct intramolecular dipole-dipole-interaction in the 13 Carbon-, Fluoro-, Phosphorus-, Deuterium- or Proton- nuclear magnetic-resonance.

Liquid crystals are particularly suited as solvents for investigating such phenomena [11, 13, 21]. Due to the parallel orientation of the molecular axes, nematic liquids are anisotropic and many of their properties are thus dependent on direction. Generally, molecules do not have a preferred orientation within large volumes. However, nematic liquids can be uniformly oriented over long ranges by magnetic fields of several thousand Gauss, by electrical fields below 5000 V/cm or in thin layers by surface interaction effects. As dissolved guest molecules are incorporated into a liquid crystal host with their long axis parallel to the preferred direction it is also possible to investigate their anisotropic molecular properties.

This work presents some examples from molecular spectroscopy, which demonstrate possibilities and limitations for use of liquid crystals as a matrix for investigating direction-dependent molecular properties.

5.1 Molecular Orientation

The orientation of molecules in a nematic liquid can be characterized by a degree of order or order parameter S which is defined as

$$S = \frac{1}{2} \langle 3 \cos^2 \Theta - 1 \rangle, \qquad 5.1$$

whereby different S-values are assigned to the three axes of a molecule, as shown for anthracene below:

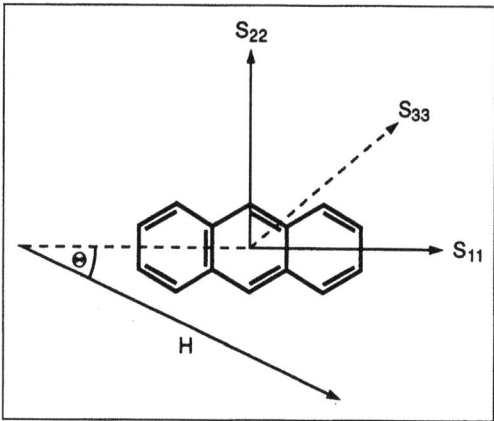

Θ is the angle between a given direction (e.g., the direction of a magnetic or electric field or perpendiculars or parallels to an interface) and the molecular axis under observation. The brackets $\langle\ \rangle$ indicate that Θ is a statistical mean of all possible Θ-values; this is because temporal fluctuations of a molecular orientation due to heat motion also occur in anisotropic solvents. Thus, a mean spatial orientation of molecules can be defined using the three S-value.

In case of non-indexed S-values S always refers to the long axis of the molecule. According to Eq. 5.1, a positive S-value indicates a preferred parallel, and a negative S-value indicates a perpendicular orientation to a given direction. The maximum value $S = +1$ indicates a completely parallel orientation of the long axes and $S = -0.5$ a completely perpendicular orientation. Accordingly, $S = 0$ applies to an isotropic phase.

It turned out that the order parameter of nematic liquid crystals has S-values between 0.2 and 0.85, whereby the S-value of the guest molecules is, as a rule, smaller than that of the host molecules.

As the thermal molecular motion counteracts parallel orientation, the S-value of the nematic solvent as well as that of the dissolved substance decreases with increasing temperature up from the melting point. At the clearing point, S drops discontinuously to zero. Generally, the temperature coefficient of the order parameter S is greater the smaller the dispersion forces, dipole interactions, and polarizabilities which act between the molecules and which are responsible for the parallel orientation, i.e., the stronger the temperature dependence of the viscosity or the smaller the mesophase range of a nematic liquid. Hence, the best orientation is obtained just above the melting point.

5.2 Selection of Appropriate Solvents

For the study of direction-dependent molecular properties, nematic liquids with a degree of order at the measuring temperature in question as high as possible and with a temperature-dependence as small as possible are well-suited as anisotropic solvents. Low melting nematic solvents with broad mesophase range and liquid crystalline at room temperature are hence the solvents of choice for spectroscopic applications.

Preferred anisotropic solvents are liquid crystals having a biphenyl-, phenylcyclohexane-, bicyclohexyl- or cyclohexanecarboxylic acid ester structure. Their absorption edges are given in Fig. 5.1. Depending on R and R', nonpolar or strongly polar anisotropic solvents can be prepared which differ similarly in their solvent properties as conventional isotropic organic solvents.

By mixing liquid cyrstals of opposite diamagnetic anisotropy, diamagnetically isotropic nematic solvents can be prepared. These are of particular use in NMR spectroscopy. In addition, by cooling rapidly to $-50°C$, mesomorphic glasses can be prepared which are important in ESR and low-temperature fluorescence and phosphorescence analyses. A list of typical liquid crystal solvents and their properties is given in Table 5.1.

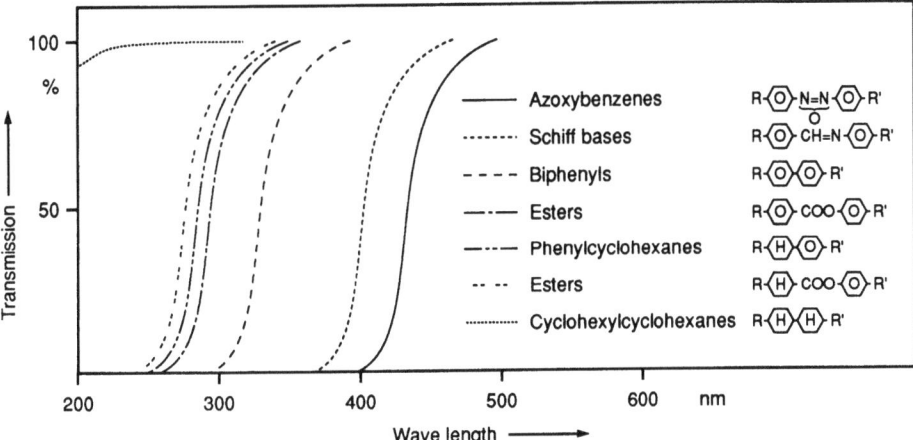

Fig. 5.1. Absorption edge of liquid crystals (layer thickness = 10 μm)

5.3 Preparation of Nematic Solutions and of Samples

A nematic solution is prepared by dissolving the substance under investigation in a liquid crystal, heating the solution to above its clearing point for better homogenization and, subsequently, cooling to the required measuring temperature.

Table 5.1. Properties of some liquid crystalline solvents used in spectroscopy.

	Eutectic / Mixture	nemat. range °C	viscosity cP / 0°C	Δn / 589 nm	$\Delta\varepsilon$ 20°C	$\bar{\varepsilon}$ 20°C
N5	CH_3O–⟨O⟩–N=N(O)–⟨O⟩–R R = C_2, C_4	−5 to +73	26	0,29	−0,2	5,9
E7	R–⟨O⟩–⟨O⟩–CN RO–⟨O⟩–⟨O⟩–CN R–⟨O⟩–⟨O⟩–⟨O⟩–CN R = C_5, C_7, C_8	−10 to +60	39	0,225	+13,8	9,8
	R–⟨O⟩–COO–⟨O⟩–CN R = $C_2, C_3, C_4, C_5, C_6, C_7$	about 0 to 50	98	0,176	+28,8	18,5
ZLI-1132	R–⟨H⟩–⟨O⟩–CN R–⟨⟩–⟨O⟩–⟨O⟩–CN R = C_3, C_5, C_7	−6 to +70	28	0,14	+10,3	8,1
	R–⟨H⟩–COO–⟨O⟩–OR' R = C_3, C_4, C_5 R' = C_1, C_2, C_4	<0 to +69	19	0,087	−1,4	4,2
ZLI-1167	R–⟨H⟩–⟨H⟩–CN R = C_3, C_5, C_7	+32 to +83	30 at 35 °C	0,06	+3,9	4,9
ZLI-1695	R–⟨H⟩–⟨H⟩–CN R = C_2, C_3, C_4, C_7	+13 to +72	61	0,063	+4,8	5,3
	60% Cholesterylchloride 40% Cholesteryloleate	20				
	66% Cholesterylchloride 34% Cholesteryllaurate	30				

Planar or elongated molecules can be dissolved in amounts of up to 30 mol% in a liquid crystal without total loss of its liquid crystalline properties. In the case of ball-shaped molecules, this limiting value is often smaller. Both melting point and clearing point of the solution (the latter more so than the former) are, however, significantly reduced. The mesomorphic range of a nematic solution is therefore always smaller than that of the pure nematic matrix.

In order to measure polarized optical spectra, samples are required whose molecules are (as far as possible) uniformly oriented over the area of the optical path involved. This is the case in thin wafers of single crystals or in stetched plastic foils. In the latter technique, the sample to be analyzed is dissolved in, for example, a polystyrene solution poured as a thin liquid film and allowed to solidify. The formed film is subsequently stretched in one direction. This process results in a limited orientation of the molecules in the direction of stretch.

The use of liquid crystalline solvents significantly facilitates the measurement of polarized spectra. To measure polarized IR spectra, for example, conventional IR recording techniques can be applied by using teflon ring spacers (10–25-μm) between two window plates (KBr, NaCl, BaF_2 wafers) to form a liquid film of defined path length. An analogous technique can be used in visible-, ultraviolet- and fluorescence-spectroscopy. Approximately 10^{-2} molar and 10- to 50-μm-thick liquid films between glass of quartz plates are required. Prior to assembling and filling the cell, however, the plates should be rubbed in one direction with a leather cloth; this causes the long axes of the liquid crystal molecules to become uniformly oriented parallel to the surface of the plates in the rubbing direction over a macroscopic range. The same orientation effect can be achieved by using a magnetic field of several thousand Gauss or by applying an electric field of up to 5000 V/cm. The direction of the applied fields depends on the diamagnetic or dielectric anisotropy of the liquid crystalline solution. As a liquid crystal layer maintains its orientation for a long period of time, measurements can be carried out without much experimental effort. Substances dissolved in a liquid crystal are oriented as the matrix, whereby the degree of order depends on the molecular structure involved. Above the melting point of the solution the order parameter ranges between $+0.3$ and $+0.4$ and is hence often larger than in stretched foils, but smaller than in single crystals.

5.4 Infrared Spectroscopy in Liquid Crystals

Normal IR spectra of nematic compounds in the liquid crystalline state are largely identical to those in the solid or isotropic liquid phase. However, they have less well defined and broader absorption bands in the spectral range of CH_2-deformation vibrations (1400–1500 cm^{-1}) and nicking vibrations (500–1000 cm^{-1}) of alkyl chains, a fact that can be attributed to differing, but limited rotational possibilities of these groups around the long axis in the nematic phase (Fig. 5.2).

Polarized IR spectra—where incident polarized light oscillates once perpendicularly and, on the other hand, parallel to the long axis of the molecule—have the same band pattern as non-polarized spectra, but show, to some extent, distinct differences in intensity, a phenomenon known as infrared dichroism. By treating the cuvette windows with, for example, lecithin or versamide, it is possible to obtain a homeotropic phase that is perpendicularly oriented to the window plates, instead of a homogeneous phase with parallel orientation to the window surface. In such cases it is possible to determine the infrared dichroism without polarized light. This is done by first recording the IR spectrum of the nematic solution followed by that of the isotropic one above its clearing point. Figure 5.3 illustrates the two orthogonally

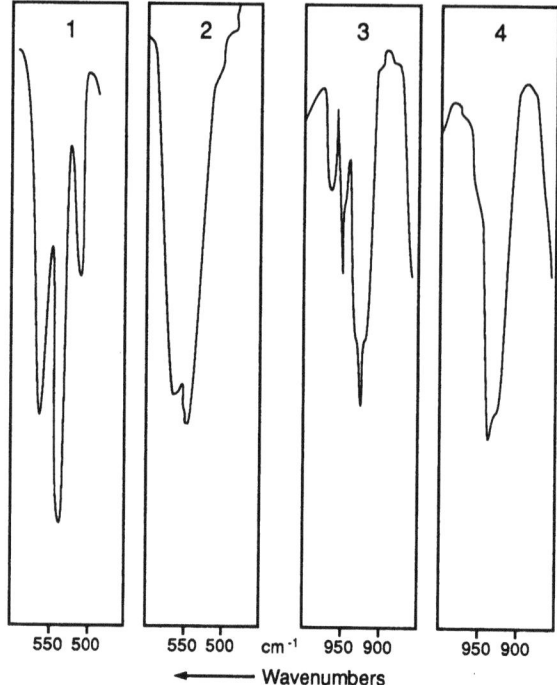

Fig. 5.2. IR spectra of crystalline (1) and nematic (2) 4,4'-Dimethoxyazoxybenzene and of crystalline (3) and nematic (4) 4,4'-Diethoxyazoxybenzene. [Accord. 1]

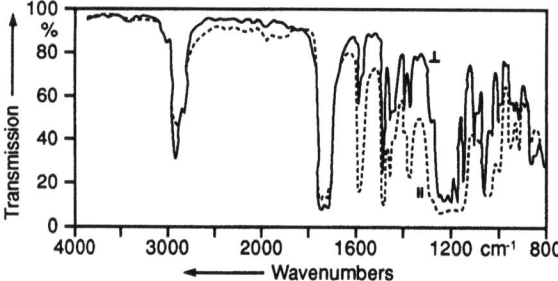

Fig. 5.3. Polarized IR spectra of 4-(4'-Ethoxyphenyloxycarbonyl) phenyl-butylcarbonate in supercooled nematic phase at room temperature. Light parallelly ($\parallel$) and perpendicularly ($\perp$) polarized to long axis of the molecule. [Accord. 3]

polarized infrared spectra of the liquid crystal p-(p'-Ethoxyphenyloxycarbonyl) phenylbutylcarbonate in the nematic phase.

Due to the great number of bands involved, nematic substances can only be limitedly applied as anisotropic solvents in IR spectroscopy. However, by selecting carefully, it is quite possible to determine the polarization of $-C\equiv N$, $-C=O$, $-C\equiv C-$ or $=C=O$ vibrations of dissolved molecules.

5.5 Determination of the Order Parameter and of the Polarization

By the above-mentioned infrared dichroism, one can determine the degree of order of a liquid crystal or that of a dissolved molecule; in addition, the polarization and, in turn, the vibration group of an infrared band can be evaluated.

In order to determine the degree of order, a band of sufficient intensity should be selected which is not superimposed by neighboring bands and which belongs to a non-degenerated vibration with known orientation of the vibration moment in respect to the long axis of the molecule involved. The degree of order S can then be easily calculated from the dichroic ratio or degree of polarization $N = E_{\|}/E_{\perp}$ with $E_{\|}$ and $E_{\perp}$ being the extinction parallel and perpendicular, respectively.

If the transition moment of the IR-band is in the long axis, the following relation holds for the order parameter S and the degree of polarization N:

$$S_{(\|)} = \frac{E_{\|} - 1}{E_{\perp} + 2} = \frac{N - 1}{N + 2} \text{ or } N_{\|} = \frac{1 + 2S}{1 - S}.$$

For a transition moment perpendicular to the long axis the analogous equation is

$$S_{(\perp)} = \frac{E_{\|} - 1}{E_{\perp} + \frac{1}{2}} = \frac{N - 1}{1 + \frac{N}{2}} \text{ or } N_{\perp} = \frac{1 - S}{\frac{S}{2} - 1}.$$

Using this method, the degree of order of, e.g., nonadiene, decadiene, and undecadiene-2, 4-acid-1 was determined, as well as the temperature dependence of the order parameter of p, p'-Dihexyloxyazoxybenzene (Fig. 5.4).

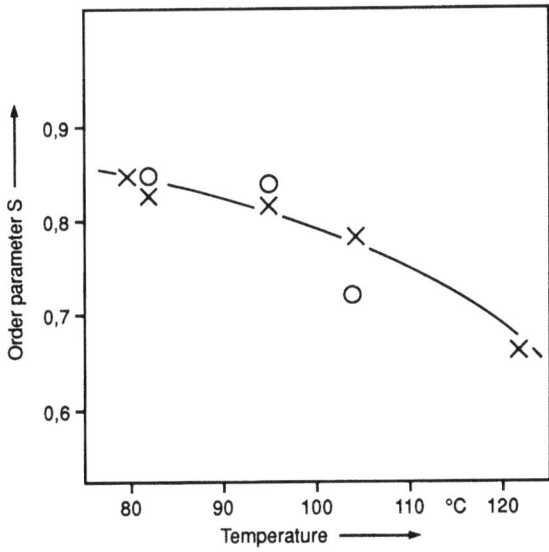

Fig. 5.4. Temperature dependence of the order parameter S of p, p'-Dihexyloxyazoxybenzene. Measurement at 780 cm^{-1} (x) and at 800 cm^{-1} (o). [Accord. 10]

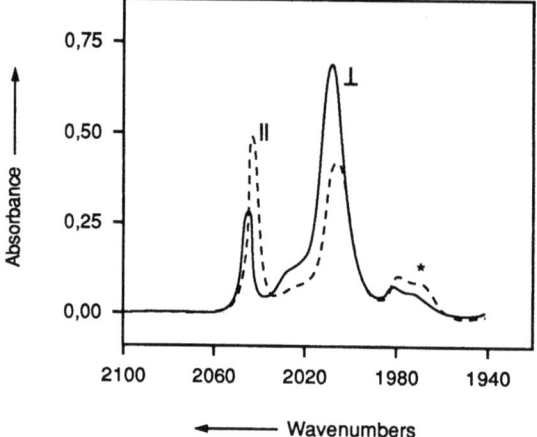

Fig. 5.5. FT-IR spectra in the $\nu(CO)$ region of $Mn_2(CO)_{10}$ dissolved in an oriented sample of Merck 1132 TNC nematic liquid crystal. $\parallel$: parallel polarization. $\perp$: perpendicular polarization. The asterisk denotes a peak due to the liquid crystal solvent. [Accord. 2]

IR dichroism was also utilized to investigate the dependence of the degree of order on the chain length of alkyloxybenzoic acids. It was observed that S increases with chain length. It has also been generally established that there is a direct dependency between the magnitude of the degree of order in respect of the long axis of the molecule and the difference between the length of the long and of the short molecular axis. Elongated molecules are hence better oriented than shorter molecules.

If, on the other hand, the spatial orientation of an oscillator in a nematic phase is known, the dichroic ratio of a specific absorption band can be used to very simply determine the polarization of the corresponding transition moment. Thus, using IR dichroism, the C=O-valence vibrations of $Mn_2(CO)_{10}$ (Fig. 5.5), $Re_2(CO)_{10}$, $Cr(CO)_6$ and $W(CO)_6$, as well as the CO-stretching vibrations of $\eta^5\text{-}C_5H_5Mn(CO)_3$ and $\eta^5\text{-}C_5H_5Re(CO)_3$ could be assigned to individual vibration classes.

5.6 Visible and Ultraviolet Spectroscopy in Liquid Crystals

The use of liquid crystals as anisotropic solvents is less limited in UV spectroscopy than in IR spectroscopy. However, due to their intrinsic absorption, liquid crystals containing a phenyl group can only be used as solvents for substances with an absorption above 300 nm.

So-called compensated cholesteric liquid crystals have frequently been used as nematic solvents in the visible and ultraviolet range which are transparent above 250 nm. Compensated cholesteric liquid crystals are binary mixtures of cholesterol esters made up of right- and left-handed cholesterol derivatives that, at a given mixture ratio and temperature, represent a nematic liquid crystals (see Table 5.1). However, they often are light scattering, which makes a correction of the spectral basis line necessary. On the other hand, when cooled to $-50°C$, they can be very easily converted into transparent, highly ordered glasses.

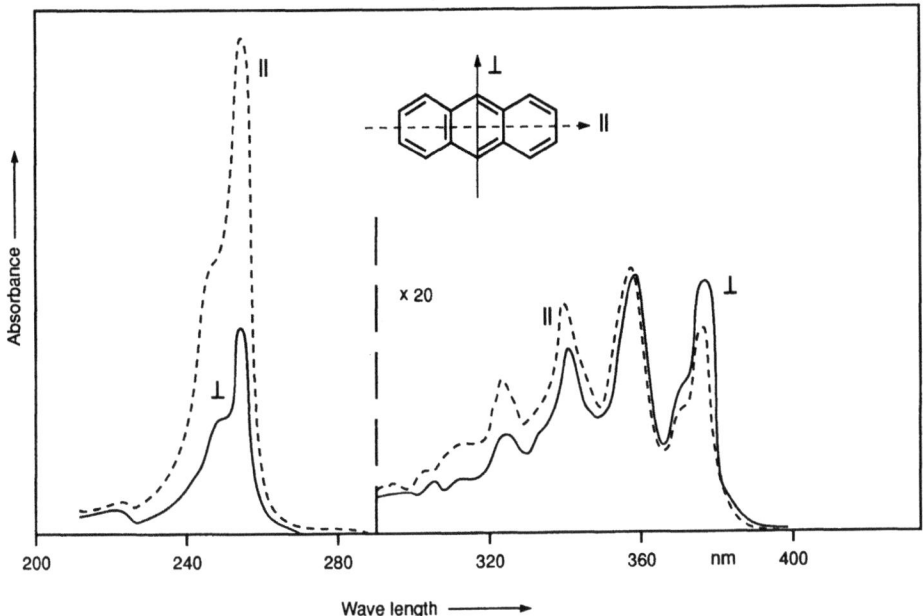

Fig. 5.6. Polarized UV absorption spectra of anthracene dissolved in the oriented nematic liquid crystal ZLl-1167 from E. Merck at 37°C. [Accord. 20]

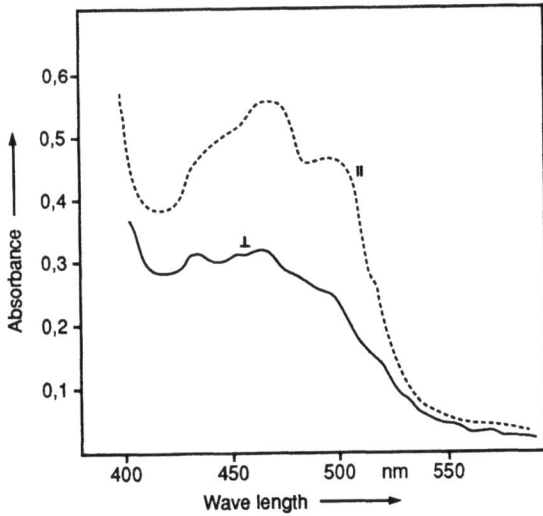

Fig. 5.7. Polarized absorption spectra of β-carotene (5×10^{-4} mole) dissolved in an oriented nematic liquid crystal (mixture of cholesteryl chloride and cholesteryl myristate, ratio 1.9:1); parallel polarization ($\|$); perpendicular polarization ($\perp$). [Accord. 15]

Liquid crystalline bicyclohexyl derivatives available today are nematic solvents that are transmissive already at 200 nm and are hence particularly suitable (Fig. 5.1). Such purely cycloaliphatic, nematic solvents open up the determination of the polarization of electronic transitions in molecules over wide spectral ranges, as shown for anthracene in Fig. 5.6. This spectrum also demonstrates that, in a molecule, various electronic transitions can be of different polarization. It can be seen that the short-wavelength transition is polarized in the long axis and the long-wavelength transition in the short axis. In β-carotene on the other hand, the long-wavelength transition is polarized in the longitudinal axis (Fig. 5.7).

5.7 Fluorescence Spectroscopy in Liquid Crystals

Polarized fluorescence spectra can be measured in nematic solvents in the same way as polarized absorption spectra. Again, the different intensity of the fluorescence

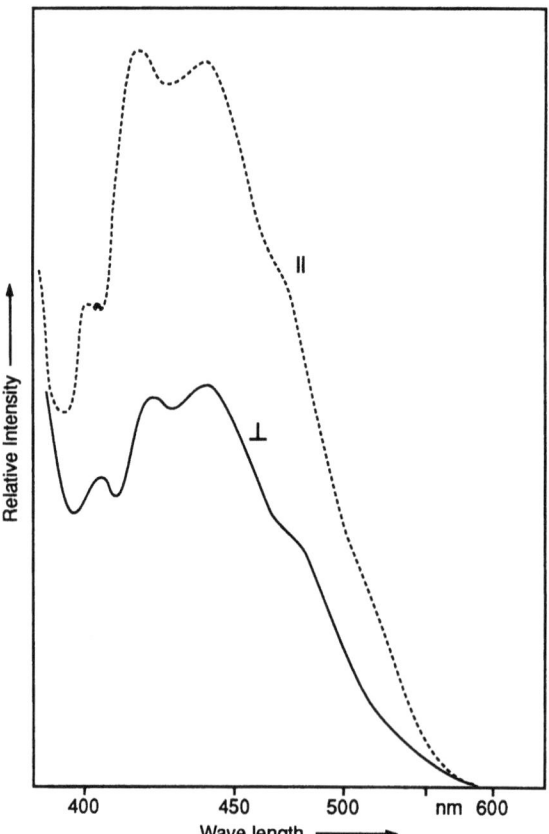

Fig. 5.8. Polarized fluorescence spectra of Diphenylhexatriene (1×10^{-4} mole) dissolved in an oriented nematic liquid crystal (mixture of cholesteryl chloride and cholesteryl laurate, ratio 1.95:1) at 30°C.; parallel polarization ($\|$); perpendicular polarization ($\perp$). [Accord. 16]

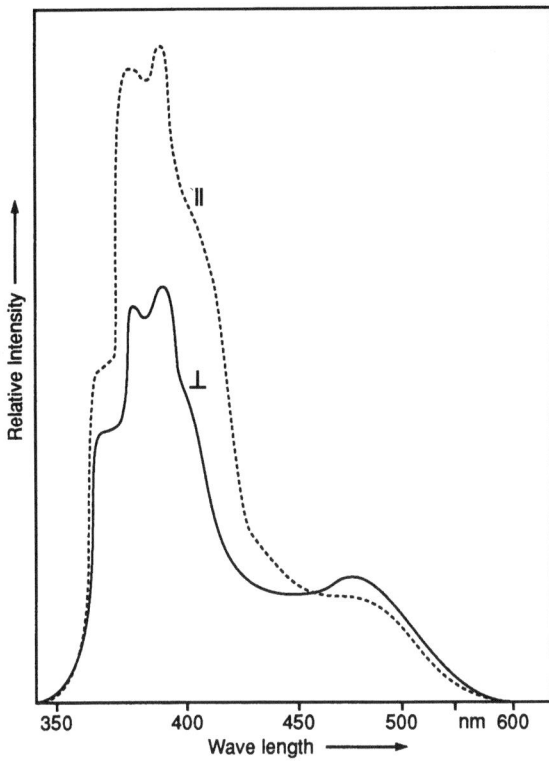

Fig. 5.9. Polarized fluorescence spectra of pyrene (7×10^{-2} mole) dissolved in an oriented nematic liquid crystal (see Fig. 5.8).

bands with light parallel and perpendicular to the long axis of the molecule can be used to determine the degree of polarization N. In fluorescence spectroscopy N is defined as the quotient of the relative intensities $I_{\parallel}$ and $I_{\perp}$. Because fluorescence of molecules is always shifted to longer wavelengths compared to absorption, nematic solvents can be applied more generally in fluorescence spectroscopy than in absorption spectroscopy.

Using diphenylhexatriene (Fig. 5.8) and pyrene (Fig. 5.9), it could be shown that the fluorescence of these compounds is polarized in the long axis. In the case of tetracene (Fig. 5.10), however, the molecule is polarized in the short axis. These few examples show that the use of nematic liquid crystals is a simpler method of recording polarized spectra than the use of stretched films or single crystal wafers.

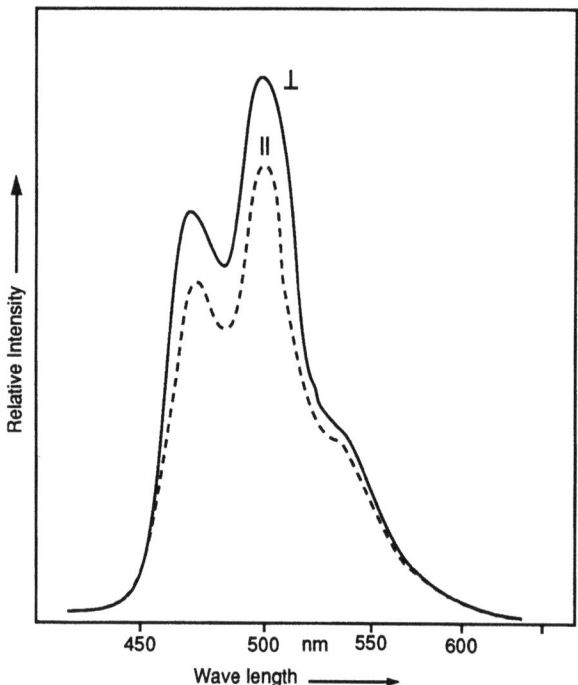

Fig. 5.10. Polarized fluorescence spectra of tetracene (1×10^{-3} mole) dissolved in an oriented nematic liquid crystal (see Fig. 5.8).

5.8 Nuclear Magnetic Resonance Spectroscopy in Liquid Crystals

Saupe and Englert have demonstrated in two basic papers that high-resolution nuclear magnetic resonance spectroscopy is also possible in nematic solvents and that this technique can be used to determine bond angles and relative bond distances, particularly in molecules with only few protons [6, 17].

In liquid crystals, just as in isotropic solutions, the intermolecular interactions of nuclear moments, e.g., the long-range and strong intermolecular direct dipole-dipole interactions are reduced to an insignificantly small mean value due to the rapid motion and diffusion of the molecules. However, the intramolecular direct dipole-dipole interactions are retained due to the fact that the long axis of the dissolved molecules are oriented preferably parallel to the long axis of the liquid crystal molecules. In the case of liquid crystals with a positive diamagnetic anisotropy $\Delta x = x_{\parallel} - x_{\perp}$, with $x_{\parallel}$ and $x_{\perp}$ being the diamagnetic susceptibility parallel and perpendicular to the director, the dissolved molecules orient themselves in the direction of an external magnetic field as well as the matrix molecules. In this way a preferred macroscopic orientation is formed.

In order to avoid averaging out to zero the intramolecular direct spin splitting which, by this macroscopic alignment becomes visible, the sample, if positioned perpendicularly to the magnetic field, should not be rotated or if so, only slowly (1 − 6 Hz); the rotation frequency in such cases is dependent on temperature and field strength. Small inhomogeneities in the sample, in the glass wall of the probe tube and in the magnetic field can, however, no longer be averaged out. The half-widths of the resonance lines thus measured, which are often sharper at the center of the spectrum than at either end, are at 3–20 Hz larger by approximately a factor of 10 than in isotropic solvents. Small vicinal or geminal indirect spin couplings can, as a rule, no longer be resolved. However, if the sample is in line with the magnetic field, which is normally the case in spectrometers with superconducting magnets, the probe tube can be rotated and substantially sharper resolved spectra with line widths of 1–5 Hz are obtained.

The same effect is to be observed by use of conventional magnets if liquid crystalline solvents with negative diamagnetic anisotropy are applied. In this case the director is aligned perpendicular to the orienting external magnetic field. Spinning of the sample produces a single, crystal-like, highly ordered nematic solution and results in a spectrum with considerably improved resolution (Fig. 5.11).

In proton NMR direct dipole-dipole couplings can amount to far in excess of 1000 Hz, in contrast to a maximum of around 20 Hz in case of indirect couplings. Therefore the spectral range encompasses 10 000 Hz and even more for small molecules compared to a few thousand Hz in isotropic solvents.

Nuclear magnetic resonance spectra in nematic solvents are thus, additionally to the chemical shift and – provided it can be resolved – to the indirect spin coupling which acts via the electron framework of the molecule, characterised by the direct spin coupling which acts through space. They are, hence, quite different from those obtained in isotropic solvents. Some simple examples from proton NMR will more clearly illustrate the principle of the dipole splitting.

Molecules with chemically and magnetically equivalent protons

Acetylene is a molecule with two chemically and magnetically equivalent protons. It shows in isotropic solvents only one resonance signal due to the fact that the Brownian motion of the molecules cancels out the dipolar coupling. If this molecule is oriented by a matrix in such a way that the long axis (in this case the straight line between the two protons) takes on a definite direction with respect to the external field, an additional field exists at the position of both protons due to the dipolar coupling. This field, depending on the spin direction of the coupling proton, differs from the original field H_0 by + or − ΔH, the contribution of the spin of the one proton at the position of the other one. Thus, the resonance of both protons is not observed at H_0, but rather at $H = H_0 \pm \Delta H$. Two signals of the same intensity appear to be centered at H_0 and are separated by $2\Delta H$ from each other. The magnitude of this line splitting corresponds to the direct spin-spin coupling constant $D_{H,H}$. The center of the doublet equals the chemical shift of acetylene in the nematic solvent (Fig. 5.12). As shielding effects in the isotropic phase are normally different

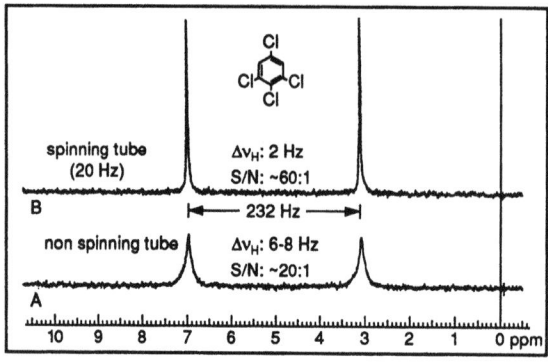

Fig. 5.11. ^{1}H-NMR spectrum of 1, 2, 3, 5 Tetrachlorobenzene (30 mole%) dissolved in a nematic liquid crystal with negative diamagnetic anisotropy (ZLI-1167, E. Merck). [Accord. 14]

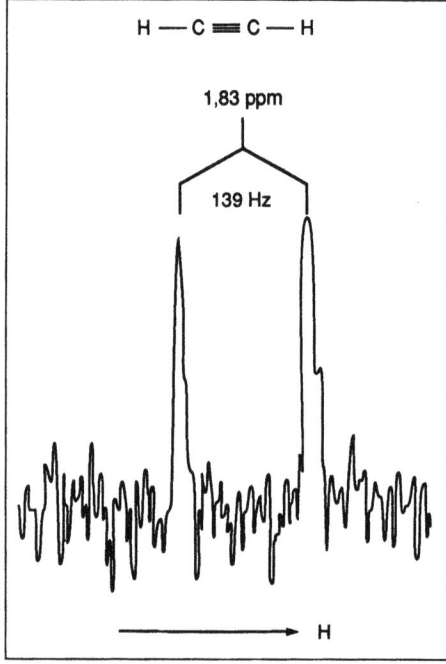

Fig. 5.12. 100 MHz ^{1}H-NMR spectrum of acetylene at 75°C dissolved and partially oriented in the nematic liquid crystal 4, 4'-Di-n-hexyloxyazoxybenzene. [Accord. 7]

from those in an anisotropic one, the chemical shift measured in isotropic and nematic solvents are often different.

Benzene is a molecule which has six chemically and magnetically equivalent protons. In isotropic solution it has, like acetylene, only one signal, but 78 lines are possible in a nematic solvent. This molecule demonstrates the limits of the method (Fig. 5.13). The numerous direct dipole couplings possible in molecules with many

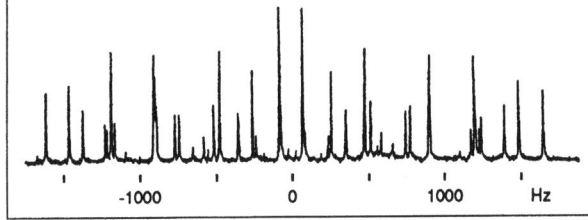

Fig. 5.13. 100 MHz ^{1}H-NMR spectrum of benzene at 50°C dissolved in a nematic liquid crystal. [Accord. 18]

protons give rise to such a splitting of resonance lines that the spectrum can only be analysed by substantial computer efforts. Often, the spectrum is hardly recognizable for noise. This is the reason why proton-rich nematic solvents themselves do not show a nuclear magnetic resonance spectrum and can thus be used as anisotropic solvents, and it is also why nuclear magnetic resonance in nematic solvents is restricted to relatively small molecules.

Molecules with chemically and magnetically non-equivalent protons

4,6-Dichloropyrimidine is a molecule that, like acetylene, has two protons but they are chemically and magnetically non-equivalent. Two singlets are seen in isotropic solvents, neglecting the very small para-coupling. In contrast, two doublets exist in nematic solution (Fig. 5.14).

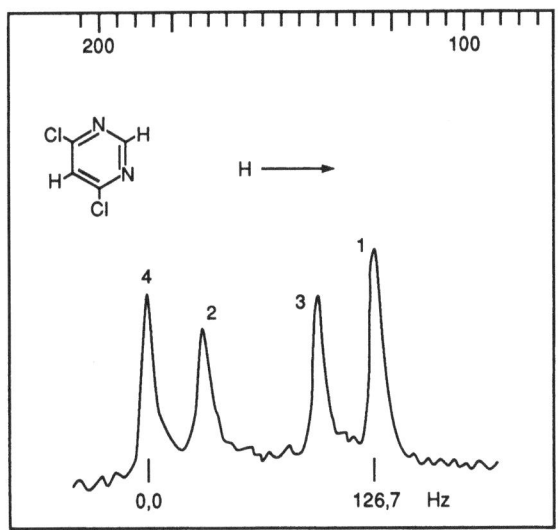

Fig. 5.14. NMR spectrum of 4,6-Dichloropyrimidine at 27°C dissolved in a nematic liquid crystal. [Accord. 4]

Bond lengths and bond angles

Direct dipole couplings can be used to determine various molecular parameters. The dipole coupling is defined by the equation

$$D_{i,j} = -\frac{h * \gamma_i * \gamma_j}{4\pi^2 * r_{i,j}^3} * \frac{1}{2} \langle 3\cos^2\Theta_{i,j} - 1 \rangle,$$

with $\Theta_{i,j}$ = the angle between direction of magnetic field and axis connecting nuclei i and j, $r_{i,j}$ = the distance between nuclei i and j and $\gamma_{i,j}$ = the gyromagnetic ratio of nuclei involved.

As the molecular orientation shows temporal fluctuations in anisotropic solvents, the brackets $\langle \ \rangle$ indicate that a statistical mean for all occurring values of Θ is involved. In the case of an isotropic molecular motion, where all orientations are equally probable, the expression $\langle 3\cos^2\Theta - 1\rangle = 0$ that means the dipolar coupling disappears. On the other hand, in the case of a completely parallel orientation of the long axes to the field, a maximum spin splitting occurs. The term in brackets represents the degree of order or order parameter S.

If the nuclei are protons, the equation is simplified to

$$D_{H_{i,j}} = -\frac{h * \gamma_H^2}{4\pi^2 * r_{H_{i,j}}^3} * S_{i,j}.$$

In the case that r is known for two protons, the coupling constant $D_{H,H}$ can be used to determine the degree of order S and, hence, the orientation of the molecule in the liquid crystal. If, on the other hand, S is known, the dipole splitting can be used to determine the relative distance between protons within a molecule.

High-field Fourier-transform nuclear magnetic resonance spectroscopy made possible NMR spectra of nuclei of low detection sensitivity and low natural abundance, e.g. ^{13}C spectra without prior isotope enrichment. The direct $^{13}C-^{1}H$–spin couplings $D_{13_C - 1_H}$ which can thus be directly obtained and which are at 1500 Hz 10 times larger than the corresponding indirect $^{13}C-^{1}H$–couplings $J_{13_C - 1_H}$ enable C–H bond lengths to be directly calculated. Via C–H-bond lengths and proton spacings also H–C–H bond angles can be evaluated. Nuclear magnetic resonance in liquid crystals is thus, along with electron diffraction and microwave spectroscopy, to a limited extent another method for the determination of bond lengths and bond angles (Table 5.2).

5.9 Electron Spin Resonance Spectroscopy in Liquid Crystals

ESR-spectra in liquid crystalline solvents are valuable tools for preparative working chemists. Liquid crystalline nematic, glassy solid state nematic, undercooled compensated cholesteric, and smectic liquid crystals can be used. In contrast to NMR-spectra, ESR-spectra in anisotropic solvents cover a frequency range similar to

Table 5.2. Bond lengths and bond angles of some simple organic molecules determined by different methods [5].

	H–C–H–angle			C–H-length in Å	
	NMR	Microwave	Electron diffraction	NMR	Electron diffraction
CH_3CN	109° 2' ± 2'	109° 16'			
CH_3OH	110° 3' ± 8'	109° 2'±45'			
CH_3I	111° 42' ± 2'	111° 25'			
Cyclobutane	108,5° ± 2°		110°	1,171	1,092
Cyclopropane	114,4°		115,1°	1,122	1,089
1,4-Cyclohexadiene	107°		103° ± 2°		
1,1-Difluoroethylene	120,6°	121,9°		1,091	
Allene	117,9°		118,5°	1,1094	1,0816

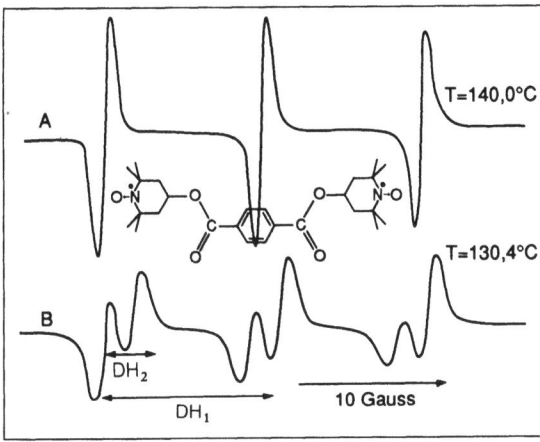

Fig. 5.15. ESR-spectra of p-Phenylene-bis[2, 2, 6, 6-tetra-methylpiperidinyl-(4)-oxycarbonyl-1-oxide] dissolved in the nematic liquid crystal p-Azoxyanisole. A: ESR-spectrum in the isotropic phase. B: ESR-spectrum in the nematic phase. [Accord. 8]

that in isotropic solvents. They only show an additional fine structure. A typical example and one that is easy to understand is the ESR-spectrum of the ground state triplet of p-Phenylene-bis[2, 2, 6, 6-tetramethyl-piperidinyl-(4)-oxycarbonyl-1-oxide], a diradical with two equivalent electron spins, measured in p-Azoxyanisole as solvent (Fig. 5.15).

The spectrum measured at 140°C in the isotropic phase consists of a triplet with a hyper-fine splitting of $H_1 = 42.9$ MHz or of about 15 Gauss. Hyper-fine splitting in ESR corresponds to indirect spin coupling in NMR. It is caused by the coupling of the unpaired electron with the magnetic moment $J = 1$ of the nitrogen. The interaction between the electron spin and the neighboring proton spin cannot be seen at the available resolution.

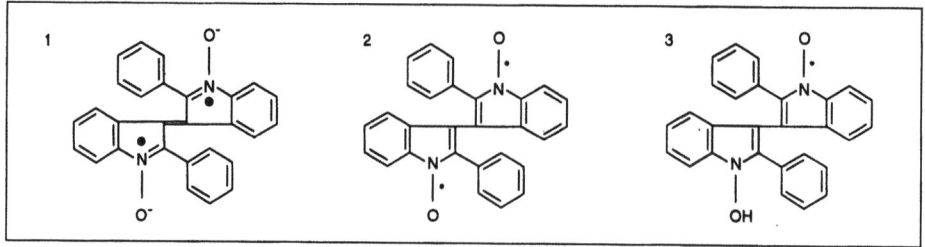

Fig. 5.16. Possible structures of Bis[2-phenylindol-(3)]-di-N-oxide. [Structure 3 holds accord. 9]

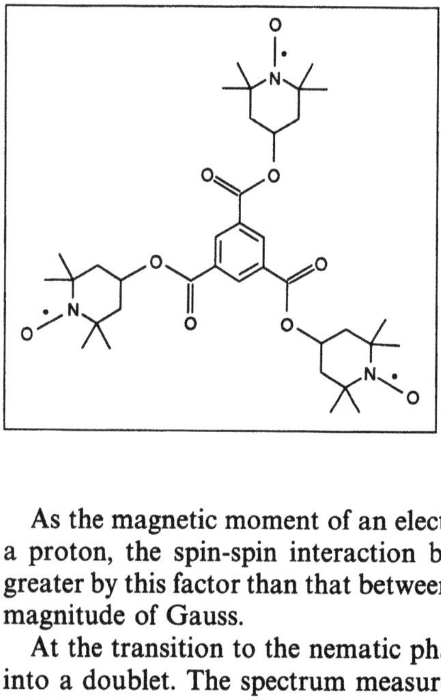

Fig. 5.17. Structure of Trimellitic acid-tris-[2, 2, 6, 6-tetramethylpiperidinol-(4) ester] tri-N-oxide. [Accord. 12]

As the magnetic moment of an electron is greater by a factor of 1825 than that of a proton, the spin-spin interaction between an electron and neighboring spins is greater by this factor than that between protons. Thus, hyper-fine splittings are in the magnitude of Gauss.

At the transition to the nematic phase each line of the triplet is split additionally into a doublet. The spectrum measured at 130°C consists of six lines. The doublet splitting ΔH_2 amounts to 8.5 MHz or approximately 3 Gauss. This additional coupling, observed in the case of oriented molecules, is caused by the dipole-dipole interaction between the two unsaturated electrons which is no longer completely averaged out. The magnitude of this splitting depends on the distance and the angle between the electron orbitals. It enables (as in nuclear magnetic resonance) information to be obtained on the order parameter S of the dissolved radicals and its temperature-dependence or on the geometric proportions between the coupling dipoles.

In addition, ESR-spectroscopy in liquid crystals makes possible (in contrast to that in isotropic solution) a definite answer on the question of whether a radical contains one or more unpaired electrons. It is thus a very simple and elegant method for distinguishing between single and multiple radicals. If the spectrum in the anisotropic phase in contrast to that in the isotropic one exhibits no additional spin splitting, the existence of a monoradical has been confirmed. An additional doublet, on the other hand, indicates the existence of a diradical, and that of a triplet indicates the existence of a triradical. In this way it was proved that, of the three possible formulae for bis[2-phenylindolyl-(3)]-di-N-oxide, only formula 3 applies (Fig. 5.16).

On the other hand, it was shown that Trimellitic acid-tris-[2, 2, 6, 6-tetramethylpiperidinol-(4) ester]-tri-N-oxide exists as a triradical (Fig. 5.17).

5.10 Mössbauer Spectroscopy in Liquid Crystals

Mössbauer spectroscopy is another field of application for liquid crystals. Due to translations of molecules being possible in all three spatial directions in a nematic phase, this phase is not a suitable matrix for Mössbauer experiments. In smectic solvents, however, in which a translation only takes place parallel but not perpendicular to the long axes within one molecular layer, a recoil-free γ-resonance is possible. In order to obtain a single-crystal-like molecular order needed for Mössbauer experiments, the smectic solution must be heated up to the nematic phase. In this phase contrary to the conditions in the smectic state it is possible to increase the degree of order by applying an electric or magnetic field or by a surface effect of the cuvette windows. The improved orientational order remains when the solution is cooled back to the smectic phase.

The line intensities of a Mössbauer spectrum in a smectic matrix depend on the angle at which the X-ray incidents related to the orientation of the long axis of the molecule, as shown for of 1, 1'-Diacetylferrocene (Fig. 5.18). The probability of a γ-resonance absorption at an X-ray incidence in the direction of the long axis of the molecule (intensity of the outside pair of lines of spectrum A) is substantially lower than in the case of an almost perpendicular incidence (spectrum B). The additional weak line in the center is caused by an iron impurity in the beryllium discs between which the smectic solution was contained.

Investigating the temperature-dependence of the intensity of Mössbauer lines, one finds that the signal intensity decreases with increasing temperature. This is due to the fact that the probability of a recoil-free absorption becomes smaller with increasing molecular motion (Fig. 5.19).

Although the intensity of Mössbauer lines is higher in a crystalline matrix than in a smectic one, as can also be seen in Fig. 5.19, measurements in a liquid crystalline solution offer the advantage of a simple sample preparation compared to the growth of single crystal wafers.

The short survey on the use of liquid crystals in molecular spectroscopy should demonstrate the easy handling, wide ranging applicability and potential of anisotropic solvents in quite different spectroscopic techniques. Despite the advantage liquid crystalline solvents offer, they have found only little attention and application in actual spectroscopic research.

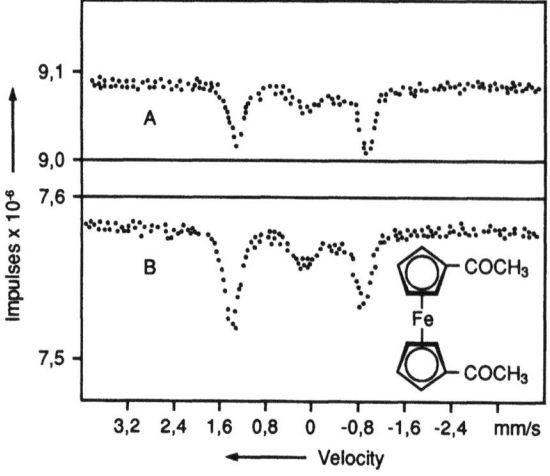

Fig. 5.18. ^{57}Fe-Mössbauer spectra of 1,1'-Diacetylferrocene (7%) dissolved in the smectic phase of 4,4'-Di-n-heptyloxy-azoxybenzene at 75°C. Angle between long axis of the liquid crystal and γ-ray: A = 0°; B = 75°. [Accord. 19]

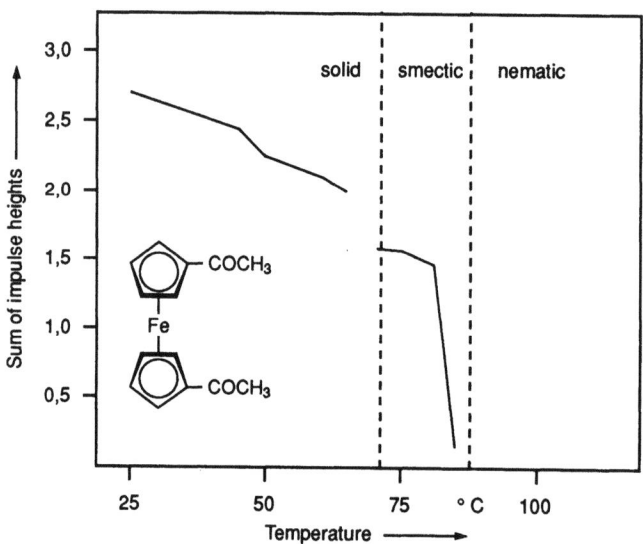

Fig. 5.19. Temperature dependence of the sum of impulse heights of the two quadrupol lines of non-oriented 1,1'-Diacetylferrocene (7%) dissolved in 4,4'-Di-n-hexyloxy-azoxy-benzene (see Fig. 5.18).

References

1. Bulkin BJ, Grunbaum D, Santoro AV (1969) Vibrational Spectra of Liquid Crystals I. Changes in the Infrared Spectrum at the Crystal-Nematic Transition. J Chem Phys 51 (4):1602
2. Butler JS, Sedman J (1988) Polarized FT-IR Spectra of Transition Metal Carbonyl Complexes Oriented in Nematic Liquid Crystal Solvents. Applied Spectr 42 (3):497
3. Caesar GP, Levenson RA, Gray HB (1969) Polarized IR Spectroscopy of Molecules Oriented in a Nematic Liquid Crystal. Application to $Mn_2(CO)_{10}$ and $Re_2(CO)_{10}$. J. Amer Chem Soc 91 (3):773
4. Diehl P (1969) NMR Grundlagen und Fortschritte, vol 1. Springer, Berlin, p 23
5. Diehl P, Khetrapal CL (1969) NMR Basic Principles and Progress, vol 1. Springer, Berlin
6. Englert G, Saupe A (1964) Hochaufgelöste Protonenresonanzspektren mit direkter magnetischer Dipol-Dipol-Wechselwirkung, Teil II. Z Naturforsch 19A:172
7. Englert G, Saupe A, Weber JP (1968) Protonenresonanzspektren orientierter Moleküle: Acetylenverbindungen. Z Naturforsch 23a:152
8. Falle HR et al (1966) The Electron Resonance of Ground State Triplets in Liquid Crystal Solution. Mol Phys 11:49
9. Forrester AR, Thomson RH, Luckhurst GR (1968) Nitroxide Radicals. Part III. The Formation of Mononitroxides from Binitrones. J Chem Soc B:1311
10. Hansen TS (1969) Liquid Crystal as a Solvent in Infrared Spectroscopy. Z Naturforsch 24a:866
11. Kelker H, Hatz R (1980) Handbook of Liquid Crystals. VCH, Weinheim
12. Luckhurst GR (1970) Electron Resonance in Anisotropic Solvents. RIC Rev 3:61
13. Michl J, Thulstrup EW (1980) Spectroscopy with polarized light. VCH, Weinheim
14. Pohl L, Eidenschink R (1978) Hochaufgelöste NMR Spektroskopie in nematischen Lösungsmitteln mit negativer diamagnetischer Anisotropie. Kontakte (Merck) 2:33
15. Sackmann E (1968) On the Polarization of Optical Transitions of Dye Molecules Oriented in an Ordered Glass Matrix. J. Amer Chem Soc 90 (13):3569
16. Sackmann E, Rehm D (1970) Fluorescence polarization measurements on molecules oriented in liquid crystals. Chem Phys Lett 4 (9):537
17. Saupe A, Englert G (1963) High-Resolution Nuclear Magnetic Resonance Spectra of Orientated Molecules. Phys Rev Lett 11 (10):462
18. Saupe A (1968) Neuere Ergebnisse auf dem Gebiet der flüssigen Kristalle. Angew Chem 80:99
19. Uhrich DL, Wilson JM, Resch WA (1970) Mössbauer Investigation of the Smectic Liquid Crystalline State. Phys Rev Lett 24 (8):355
20. Wedel H, Haase W (1978) Substituted Phenylcyclohexanes and Cyclohexylcyclohexanes. Two New Classes of Liquid Crystals as Anisotropic Solvents in Optical Absorption Spectroscopy. Chem Phys Lett 55 (1):96
21. Bahadur B (1991) Liquid Crystals Applications and Uses, vol. 2. World Scientific

6 Liquid Crystal Displays

M. Schadt

Introduction

There have been three major prerequisites for the establishment and the growing success of the liquid crystal display (LCD) technology since the early 1970s. Namely, the discovery of electro-optical field-effects on which today's LCDs are based, the successful search for liquid-crystal materials with suitable material properties, and last but not least, the development of the technological tools which are required for manufacturing LCDs.

To establish and to improve the increasingly complex interaction between man and machine requires sophisticated electro-optical devices. These devices must be compatible with highly integrated and low power consuming electronics. Moreover, displays, which should preferably be flat and of low weight, must be capable of rapidly transforming the electronic output and input signals of a multitude of electronic equipment into high quality optical images. Since the LCD technology is the only existing display technology which is capable of displaying equally well the simple time information of a digital wrist watch as well as the highly complex color images of future high-definition television sets, it will become a key technology in the years to come.

With the advancement of the LCD technology, spin-offs have been created which open up new applications within the field, as well as in non-display-related new areas. Examples for spin-offs are the development of the thin-film transistor technology which is required for actively addressing high information content twisted nematic (TN) LCDs for television and computers, or the design of novel, non-centrosymmetrically ordered organic materials for integrated optical applications. For a recent review of the development of the LCD and LC material technologies [1,2].

To illustrate the rapid progress of the LCD and the LC materials technologies it is interesting to note the simplicity of the first commercial twisted nematic liquid crystal mixture which was developed by Schadt et al. in 1971 after the twisted nematic effect was discovered. The mixture consisted of just two LC components (Schiff's bases) with almost identical molecular structures [3]. Despite its simplicity, its properties were sufficient to realize the first 3.5 digit commercial prototype TN-LCD depicted in Fig. 6.1 which comprised 24 picture elements (pixels). For comparison, today's most advanced LC materials, which are used in thin-film transistor (TFT) addressed color television TN-LCDs as well as in supertwisted nematic (STN) LCDs for computer terminals, frequently consist of more than 20 LC components with very different and complex molecular structures. The complexity of modern LC materials is a consequence of the many, often contradictory material requirements which have to be designed into them to meet the desired LCD performance, e.g., short response times,

Fig. 6.1. First 3.5 digit prototype TN-LCD made in 1971 at F. Hoffmann-La Roche.

broad operating temperature ranges, low residual ionic impurities, steep voltage-transmission characteristics for high multiplexibility (cf. Section 6.1.3) etc. Most of the LC components used in today's mixtures for high information content LCDs have been designed only during the past few years. From combining these modern LC materials with today's most advanced actively addressed TN-LCDs there result color displays with more than one million pixels. This is an increase in the number of pixels by more than a factor of 40 000 since the first 3.5, black-and-white TN-LCDs in 1971.

The demand for large-area, direct-view and projection LCDs capable of displaying the high information content of computer terminals and television screens not only requires new liquid crystals and actively addressed display substrates, it also spurs the search for new electro-optical effects with steep transmission-voltage characteristics, such as those resulting from various supertwisted nematic configurations. Such configurations have proven to be highly multiplexible thus drastically reducing the number of pixel connections and integrated drivers.

The versatility of LCDs allows their use, not only as direct-view displays, but also as light modulators in large-area projection systems. Since state of the art, actively addressed direct-view TN-LCDs are still restricted in size to 14-inch diagonals, LCD projection systems are interesting alternatives for displaying large-area television pictures. This can be achieved by front or by rear projection. Because the realization of future high-definition television depends crucially on the progress that will be made in flat and low weight, high information content displays, the realization of compact and bright LCD projection systems will have a strong impact on this development.

Virtually all of today's commercial LCDs are based on the twisted nematic or on supertwisted nematic effects, that is, they all comprise nematic liquid crystals. However, because of the interesting properties of ferroelectric smectic C (S_C^*) liquid crystals (cf. Section 6.2) with their short response times, optical bistability and/or gray scale capability, LCDs need not remain an exclusive nematic domain.

6.1 Electrooptical Field Effects for Nematic LCDs

6.1.1 Liquid Crystal Field Effects in General

A prerequisite to render liquid crystals applicable in displays are electro-optical effects based on which their optical appearance can be modulated by applying an electric field. The effect should have the potential to be broadly applicable, to be manufacturable, and to pave the route to new applications and devices. The properties of the effect and the possibility to design liquid crystals which meet its requirements determine the usefulness of the effect. Different applications may require LCDs that are based on different electro-optical effects and/or different LC materials. Moreover, the LCDs have to be competitive with alternative technologies, as for instance with light emitting diodes (LEDs), vacuum fluorescent displays (VFDs) or with the high information content color cathode ray tubes (CRTs). Therefore, the effect has to offer technological advantages such as large contrast between off- and on-state, a wide range of view, full gray scale and color reproducibility, large design flexibility, compatibility with low-power and low-voltage integrated driving circuits, flat design, low weight, reliability, operability in transmission as well as in reflection and – most important for future complex man-machine interfaces: its electro-optical characteristics should have the potential to reproduce full color images with high information content such as those of computer, television and high definition TV screens.

Since the beginning of today's nematic field effect LCD technology some 20 years ago, a remarkable number of potentially applicable liquid crystal field-effects was discovered. They are chronologically listed in Table 6.1. Up to now, the electro-optical effects almost exclusively used in LCDs are the twisted nematic (TN) effect by Schadt and Helfrich [4] and various supertwisted nematic (STN) effects [5–10] (Table 6.1). The work on STN effects in the 1980s was essentially spurred by the discovery of the superbirefringence (SBF) effect by Scheffer and Nehring [6]. Because of their steep transmission-voltage characteristics supertwisted configurations exhibit a strongly improved time multiplexibility. Thus, STN-LCDs do not require actively addressed LCD substrates to drastically reduce the number of connections to the LCD. However, because of their highly twisted nematic helices which lead to the steep transmission-voltage characteristics of STN-LCDs, also slower response times and a limited number of gray levels result compared with TN-LCDs. Thus, STN-LCDs are not yet suitable for television screens, however, they are effectively used in computer terminals. Figure 6.2 shows a black-white STN-LCD for a laptop computer.

Table 6.1. Liquid crystal field effects discovered since the twisted nematic effect.

Year	Abbr.	Name	Reference
1971	TN-LCD	Twisted Nematic	Schadt and Helfrich, Appl. Phys. Lett. 18: 127.
1971	DAP-LCD	Deformation Aufgerichteter Phasen	Schiekel and Fahrenschon, Appl. Phys. Lett. 19: 391.
1974	GH-LCD	Guest-Host	White and Taylor, J. Appl. Phys. 45, 4718.
1983	GH-STN	GH-Supertwisted Nematic	Waters et al, SID Proc. Int. Display Conf. Japan, 396.
1984	SBE-LCD	Superbirefringence Twisted Nematic	Scheffer and Nehring, Appl. Phys. Lett. 45, 1021.
1985	STN-LCD	Supertwisted Nematic	Kando, Nakagawa and Hasegawa, German Patent DE 3403259
1987	OMI-LCD	Optical Mode Interference	Schadt and Leenhouts, Appl. Phys. Lett. 50, 236.
1987	DSTN-LCD	Double Layer Supertwisted Nematic	Katoh et al., Japn. J. Appl. Phys. 26, L1784.
1987	BW-STN-LCDs	Film Compensated (Black-White) Supertwisted Nematic	O. Okumura et al, ITEJ Tech. Report 11, 27.
1986	PELCs	Polymer Encapsulated LCs	Doane et al, Appl. Phys. Lett. 48, 269. Fergason, SID 16, 88.
1989			T. Fujisawa et al, Proc. Jpn. Display '89, 690.
1980	SSFLC	Surface Stabilized Ferroelectrics	Clark and Lagerwall, Appl. Phys. Lett. 36, 899.
1988	DHF	Deformed Helical Ferroelectrics	Beresnev, Chigrinov, Dergachev, Pozhiadev, Fünfschilling and Schadt, Liquid Crystals 5, 1171.
1988	TSF	Tristable Switching Ferroelectrics	Chandani, Hagiwara, Suzuki, Ouchi, Takezoe and Fukuda, Jpn. J. Appl. Phys. 27, L729
1990	SBF	Shortpitch Bistable Ferroelectrics	Fünfschilling and Schadt, Digest SID, 106.
1992		Twisted S_C^*	Patel, Appl. Phys. Lett. 60, 280.

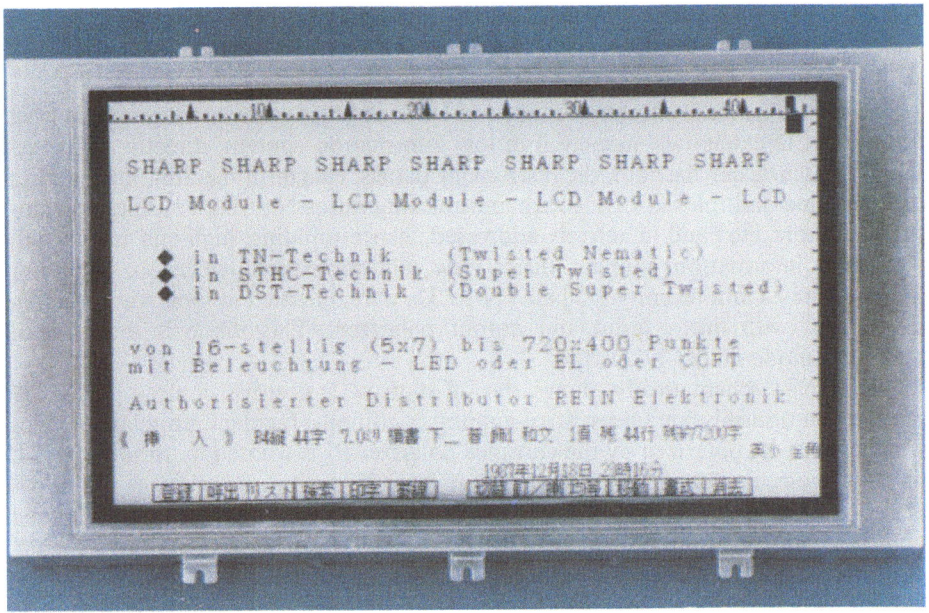

Fig. 6.2. Film-compensated, black-white supertwisted nematic LCD (BW-STN-LCD).

Recently, polymer encapsulated liquid crystals [11–13] (Table 6.1) have attracted much interest for large-area window applications as well as for all those applications where the electrically tunable light-scattering of the effects lead to large optical contrasts and low light losses (no polarizers required). If hysteresis-free transmission-voltage characteristics [13] and large material resistivities are achieved, actively addressed polymer encapsulated LC projection devices [14] with high information content become feasible.

Despite the great efforts that have been made to manufacture ferroelectric LCDs since the discovery of ferroelectricity in smectic C* liquid crystals by Meyer and coworkers in 1975 [15], and since the discovery of the surface-stabilized ferroelectric liquid crystal (SSFLC) effect by Clark and Lagerwall [16] in 1980 (Table 6.1), a commercial breakthrough has not yet been made. In view of the interesting features of SSFLCs, such as optical bistability and short response times in the 50 μs range at room temperature, this is regrettable. However, because of the complexity of ferroelectric liquid crystals the prospects of finding other field-effects which allow to make practical use of the interesting properties of ferroelectric LCs seem good (Section 6.2)

In the following the operating principles and the LC material requirements of TN-and STN-LCDs and their recent development are reviewed.

6.1.2 Wave-Guiding Operation of Twisted Nematic Liquid Crystal Displays (TN-LCDs)

The twisted nematic effect by Schadt and Helfrich [4] is today predominantly used in three LCD categories, namely i) in low information content, directly addressed LCDs (watches, car dashboards, instrument panels, etc.),in ii) medium information content, time-multiplexed LCDs with multiplexing ratios $N \lesssim 64:1^{*)}$ (office machines, calculators, etc.) and in actively addressed, fast responding, high-end LCDs with very high information content (direct-view and projection television, computer graphics, etc.). Until recently [17] all TN-LCDs were operated in their wave-guiding modes, i.e., with linear input and output polarizers. This mode is the original operation mode of TN-LCDs [4].

Figure 6.3a schematically shows the off-state of a positive contrast TN-LCD operated in transmission under wave-guiding conditions. Incident, unpolarized light which is linearly polarized by the entrance polarizer P_2 is rotated by the 90° twisted nematic molecular configuration such that its polarization direction at the upper glass substrate parallels that of the (crossed) exit polarizer P_1. Thus, in the crossed polarizer configuration of Fig. 6.3 the TN-LCD is transmissive in the off-state. A prerequisite for optimal wave-guiding of linearly polarized light to occur in the twisted nematic helix is [18, 19]:

$$\delta = \Delta n d / 2\lambda = 0.5, 1, 1.5, \ldots, \tag{6.1}$$

where δ is the optical retardation which the birefringence Δn of the liquid crystal induces over the cell gap d of the TN-LCD; λ = wavelength of incident light.

The voltage-induced change of transmission between off- and on-state of a TN-LCD results from the combination of the twisted nematic molecular configuration imposed by the LCD boundaries (Fig. 6.3a) and from specific liquid crystal material parameters. Crucial LC material parameters are the birefringence which has to meet Eq. 6.1, the positive dielectric anisotropy $\Delta\varepsilon > 0$ and appropriate elastic restoring forces. Because of $\Delta\varepsilon > 0$ the nematic director in the center of the twisted nematic configuration of Fig. 6.3a gradually aligns parallel to the electric field when applying a voltage $V > V_c$ to the TN-LCD, that is, perpendicular to the transparent electrode surfaces. The field-induced realignment of the molecules requires, depending on the LC material, 1 – 3 volts. Realignment only starts above a threshold voltage V_c above which the elastic restoring forces that stabilize the helical off-state configuration are compensated for by the torque which the electric field exerts on the helix. At voltages far above mechanical threshold for helix deformation, the director alignment becomes virtually perpendicular to the electrode surfaces (Fig. 6.3b). An optically uniaxial nematic layer results which no longer acts as a wave-guide. Thus, the crossed polarizer configuration in Fig. 6.3 is non-transmissive in the on-state (positive contrast configuration). Aligning the two polarizers in Fig. 6.3 parallel reverses the contrast of the TN-LCD from positive to negative, that is, the off-state becomes non-transmissive and the on-state transmissive. Instead of in transmission TN-LCDs are also operable in reflection.

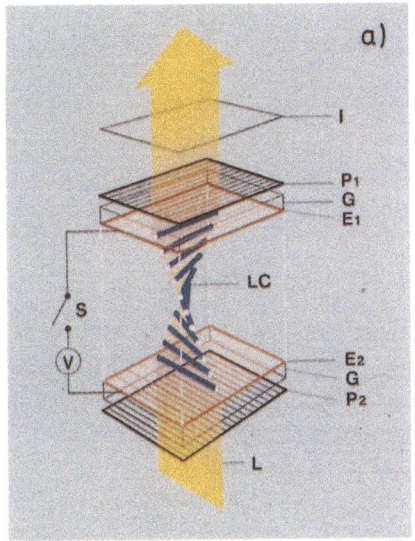 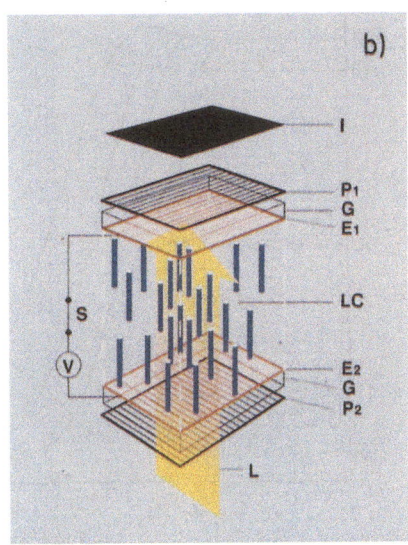

Fig. 6.3. a) Off-state configuration of a positive contrast twisted nematic (TN)-LCD. b) On-state. P_1, P_2 = linear polarizers; G = glass substrates, E_1, E_2 transparent electrodes coated with aligning layers; LC = twisted nematic liquid crystal configuration.

Figure 6.4 shows the static transmission-voltage characteristics of a positive contrast TN-LCD for different angles θ of incident light; $\theta = 0°$ = vertical light incidence. The angular dependence of the transmission results from the different optical retardation which occurs when changing the angle between incident light and TN-helix. For vertical light incidence ($\theta = 0°$) Schadt and Gerber have shown that the optical threshold voltage V_{90} for 90% transmission of a positive contrast TN-LCD operated in its second transmission minimum ($\Delta nd/2\lambda = 1$ in Eq. 6.1) can be approximated by a rather simple analytical expression [20]

$$V_{90} \propto V_c \left[2.04 - \frac{1.04}{1 + k_3/k_1} \right] \left\{ 1 + 0.12 \left[\left(\frac{\Delta\varepsilon}{\varepsilon\perp} \right)^{0.6} - 1 \right] \right\}, \qquad 6.2$$

where the threshold voltage V_c for mechanical deformation of the TN-helix is [4]

$$V_c = \pi\{[k_1 + (k_3 - 2k_2)/4]/\varepsilon_0\Delta\varepsilon\}^{1/2} = \pi[\kappa/\varepsilon_0\Delta\varepsilon]^{1/2}. \qquad 6.3$$

From Eqs. 6.2 and 6.3 it follows that both the optical threshold in the wave-guiding mode of a TN-LCD and its mechanical threshold $V_c < V_{90}$ depend on the dielectric anisotropy $\Delta\varepsilon$ and on the splay (k_1), twist (k_2) and bend (k_3) elastic constants of the LC material. Large dielectric anisotropies and/or small elastic expressions $\kappa = [k_1 + (k_3 - 2k_2)/4]$ lead to low thresholds and therefore to low operating voltages.

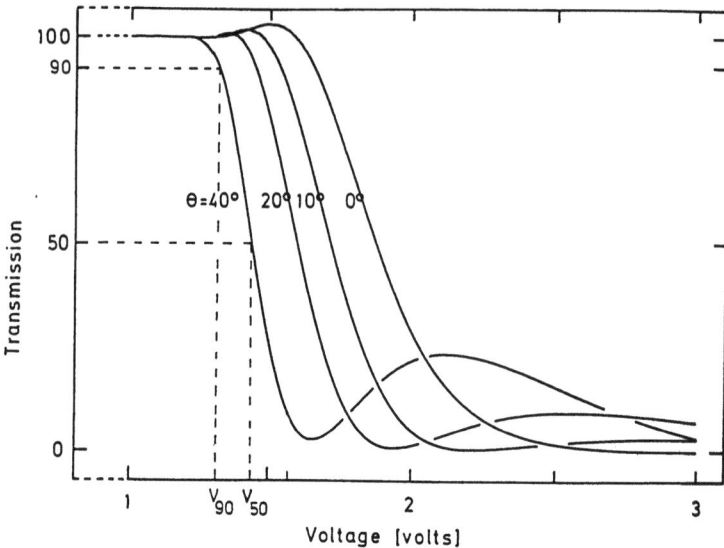

Fig. 6.4. Static transmission-voltage characteristics of a positive contrast TN-LCD operated in its wave-guiding mode at different angles θ of light incidence; vertical light incidence $\theta = 0°$.

Like the threshold voltage, also the slope of the voltage-transmission characteristics of TN-LCDs (Fig 6.4), which can be defined by the slope parameter

$$m = (V_{50} - V_{90})/V_{90}, \qquad 6.4$$

depends on LC material properties. Small m-values, that is, steep characteristics result from low bend/splay elastic ratios k_3/k_1; whereas flat characteristics which allow to display a large number of gray levels require large k_3/k_1 ratios [20]. Again, in the second transmission minimum and at vertical light incidence the slope parameter of a TN-LCD can be approximated by [20]:

$$m = 0.133 + 0.0266 \left(\frac{k_3}{k_1} - 1\right) + 0.443 \left(\ln\frac{\Delta nd}{2\lambda}\right)^2. \qquad 6.5$$

We shall see below that flat characteristics are required in high quality, actively addressed television TN-LCDs, whereas steep characteristics with a consequently limited number of gray levels are a prerequisite to achieve large multiplexing ratios in time-multiplexed LCDs that exhibit no inherent memory. Apart from LC material parameters the slope also depends on whether the TN-LCD is operated in its first or in its second transmission minimum (Eq. 6.1, [19]). Operation in the first minimum reduces the slope as well as V_{90} [21, 22].

Not only the static performance of TN-LCDs but also their dynamics depend crucially on LC material properties. In a small-angle approximation Jakeman and Raynes [23] have shown that the turn-on (t_{on}) and the turn-off (t_{off}) times can be approximated by

$$t_{on} \propto \eta d^2/(\Delta\varepsilon E - k\pi^2); \quad t_{off} \propto \eta d^2/k\pi^2, \qquad 6.6$$

where k was shown [24] to correspond to the elastic expression κ in Eq. (3), whereas the viscosity η corresponds to the rotational viscosity γ_1 [25]; E = applied electric field and d = LCD cell gap. Exact solutions of the dynamics of TN-LCDs were derived by Berreman [26] and by Van Doorn [27]. From the approximations (6.6) and Ref. [24, 25] it follows that short response times not only require LC materials with low rotational viscosities γ_1 and large elastic expressions κ, but also small cell gaps d. By introducing C = C-double bonds at specific positions in the side chains of liquid crystals Schadt, Buchecker and Müller have shown that the elastic properties of liquid crystals can be tuned over a wide range [28]. As an additional advantage, very low visco-elastic ratios which lead to short response times in TN-LCDs can be achieved with alkenyl LCs [28].

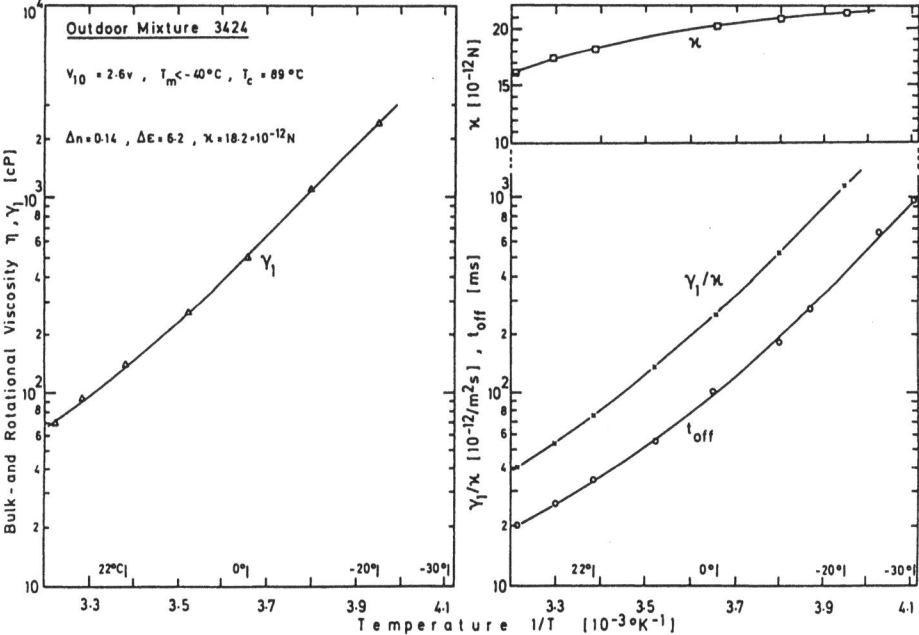

Fig. 6.5. Temperature dependencies of the rotational viscosity γ_1, the elastic expression $\kappa = [k_1 + (k_3 - 2k_2)/4]$, the visco-elastic ratio γ_1/κ and t_{off} of the early, broad temperature range liquid crystal mixture 3424 from F. Hoffmann-La Roche. t_{off} = turn-off time of a TN-LCD comprising 3424.

Figure 6.5 shows the temperature dependencies of the LC material parameters γ_1, κ, γ_1/κ as well as the corresponding TN-LCD response time t_{off} of an early, broad temperature range LC mixture form F. Hoffman-La Roche. From Fig. 6.5 follows that the strong temperature dependency of t_{off} is mainly due to γ_1 (T). As suggested by Eq. 6.6 the temperature dependence of t_{off} parallels that of its visco-elastic ratio $\gamma_1(T)/\kappa(T)$ (Fig. 6.5).

6.1.3 Time Multiplexing of LCDs

With increasing number of picture elements (pixels) in an LCD the number of connections eventually becomes so excessively large that addressing each pixel via individual connections is no longer feasible. One way to overcome this problem of directly addressed LCDs which possess no inherent optical memory is to make use of the steep transmission-voltage characteristics and their well defined threshold voltage (Fig. 6.4). As was shown by Alt and Pleshko [29], both properties are prerequisites for time multiplexing a matrix-type array of (transparent) LCD electrodes whose overlapping areas form the pixels.

Figure 6.6 schematically shows a time-multiplexed matrix array of LCD electrodes consisting of columns C on the lower glass substrate and lines L on the upper substrate. The gap between the two substrates comprises the LC molecules whose configuration is determined by the electro-optical effect on which the LCD is based. In the case of a positive contrast TN-LCD, entrance and exit polarizers are crossed and on segments appear black in the image plane I (Fig. 6.6). Depending on the optical pattern to be generated, different sequences of voltage pulses with different amplitudes have to be applied to the LCD matrix. In Fig. 6.6 the normalized pulses are chosen such that they add up to potential differences of either 1 or 3 across the matrix intersections. Then, from the TN characteristics at $\theta = 0°$ in Fig. 6.4 follows that no optical signal is generated at vertical light incidence if the voltage difference between rows and columns $V_{ns} = |1| < V_{90}$, whereas the superposition of $V_s = 2 > V_{90}$ and $V_{ns} = -1$ leads to a potential difference $V = |3| \gg V_{90}$ across the pixel. Thus, applying the time-multiplexed pulse sequence depicted on the left of Fig. 6.6 to the LCD matrix creates the optical pattern depicted in the image plane I of Fig. 6.6. Any other pattern can be generated by appropriate pulse sequences. The example shows that multiplexing allows to significantly reduce the number of connections of a LCD matrix with N lines and M columns from $(N \times M) \to (N + M)$.

The maximum optical contrast between off- and on-intersections of a multiplexed LCD depends on the slope of its transmission-voltage characteristics and on the number of multiplexed lines N. This follows from [29]:

$$N = [(1 + m)^2 + 1]^2/[(1 + m)^2 - 1]^2, \qquad 6.7$$

where m is the slope parameter of the voltage-transmission characteristics of the LCD. In the definition of Eq. 6.4, m corresponds to a contrast of $V_{50}/V_{90} = 5:1$.

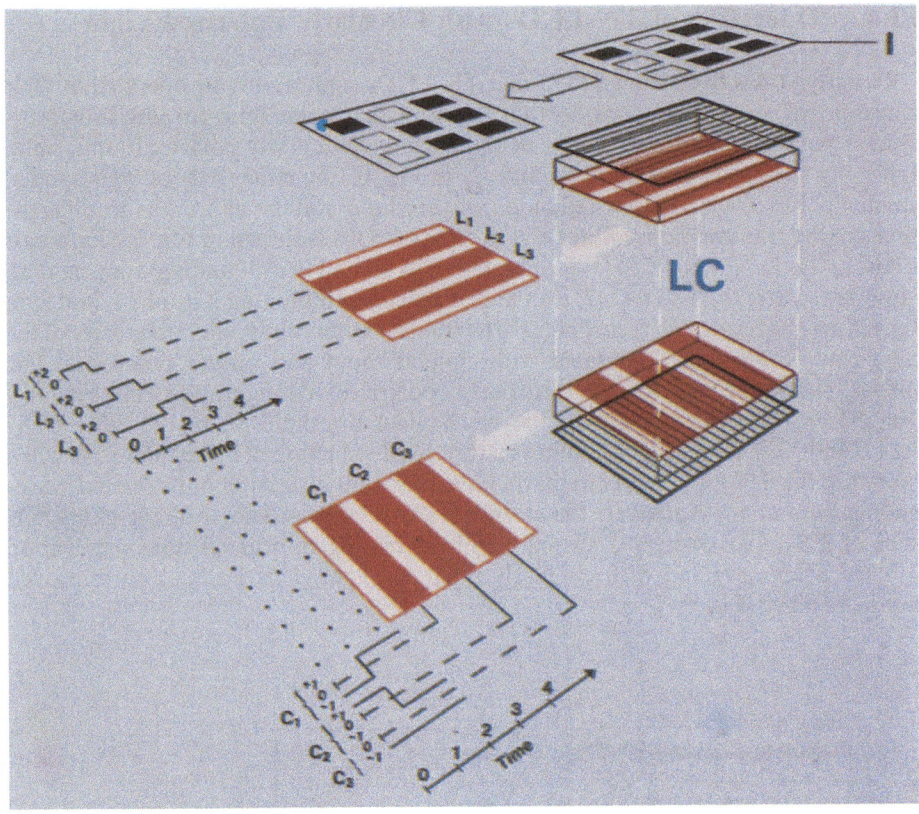

Fig. 6.6. Principle of time multiplexing a matrix array of lines L and columns C of a TN-LCD. I = image plane.

From Eq. 6.2 and 6.5 and from Fig. 6.4 it follows that the number of lines that can be multiplexed in a TN-LCD not only strongly depends on the dielectric, optical and elastic LC material parameters, but also on the angular dependence of its transmission-voltage characteristics. Since the select and the non-select voltages are determined by N and by the desired contrast, any angular dependence of the characteristics reduces N and/or the optical quality of the LCD over its field of view. Thus, prerequisites for high multiplexibility over a broad range of view are LCD characteristics with well defined and angular independent optical thresholds as well as steep characteristics. $m = 0$ corresponds to a step function which theoretically allows to multiplex an infinite number of lines with a contrast as large as that of a directly addressed LCD; Eq. 6.7.

6.1.4 Operation of TN-LCDs with Circularly Polarized Light

Recently, it has been shown [17] that TN-LCDs cannot only be operated in their conventional wave-guiding mode, that is, with linear polarizers, but also in various circular polarized modes with left- or right-handed circularly polarized input light. Replacing the linear polarizers P_1 and P_2 in Fig. 6.3 by either left- or righthanded circular polarizers leads to a reduction of threshold voltage as well as to different slopes of the transmission-voltage characteristics. This is shown in Fig. 6.7 for a pair of TN-LCDs [17]. TN-LCD (1) is operated in its second (wave-guiding) transmission minimum, whereas TN-LCD (2) is operated in its first minimum. Graphs 1' and 2' in Fig. 6.7 are the respective recordings of the transmission-voltage characteristics of the same pair of TN-LCDs operated with circular input and output polarizers. The optical threshold V_o of the circularly polarized configuration in Fig. 6.7 was shown to be identical with the mechanical threshold V_c of the TN-LCD; Eq. 6.3 and [33]. From Fig. 6.7 follows a threshold reduction of more than 30% when operating the TN-LCD with circularly polarized light instead of in its second wave-guiding minimum. Moreover, the steepness of the transmission-voltage characteristics of TN-LCDs increases strongly with increasing optical off-state retardation $\delta = \Delta nd$ when operated with circularly polarized light. This leads to an improved multiplexibility (Fig. 6.7).

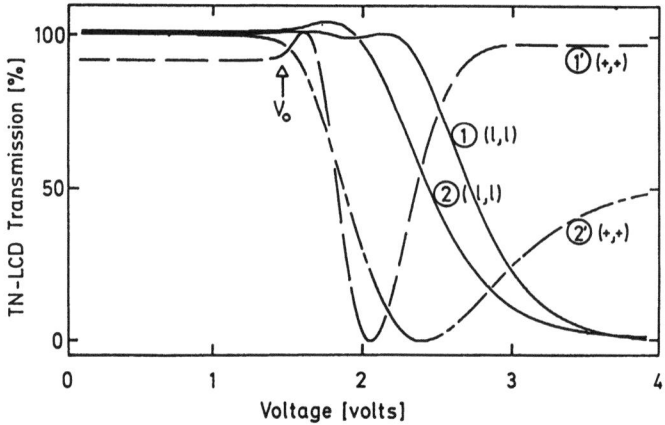

Fig. 6.7. Transmission-voltage characteristics of two TN-LCDs which are operated in their wave-guiding modes (graphs 1 and 2) as well as in their circularly polarized modes (graphs 1' and 2'). Their respective optical retardations are: δ(TN-LCD 1) = 1 and δ(TN-LCD 1) = 0.5.

6.1.5 Supertwisted Nematic LCDs and LC Materials

Because of their rather flat transmission-voltage characteristics which result even for LC materials with very low bend/splay elastic ratios $k_3/k_1 = 0.5$ (Eq. 6.5 and [28])

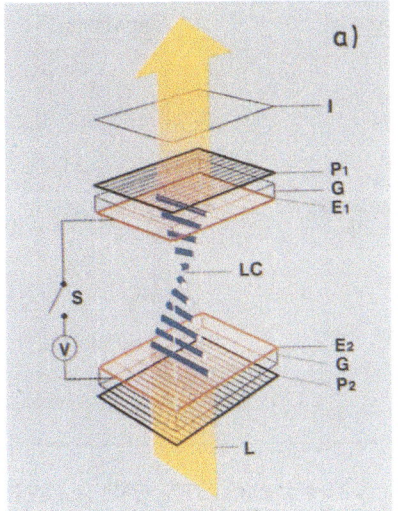 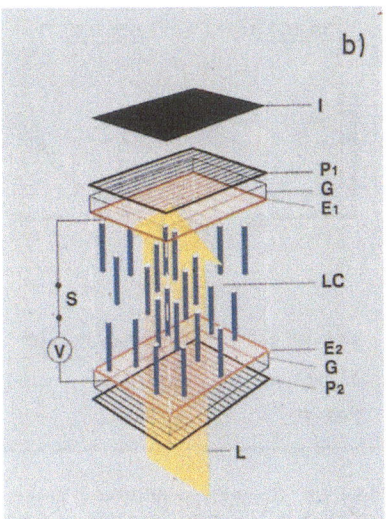

Fig. 6.8. Off-state (a) and on-state (b) of a 180° twisted, positive contrast supertwisted nematic LCD.

the multiplexibility of conventionally operated TN-LCDs is limited to about 100 lines. To overcome this limitation there are essentially three quite different approaches which can be pursued, namely a) addressing each pixel via in-situ drivers, or b) use of electro-optical effects with inherently steep transmission-voltage characteristics and therefore improved multiplexibility, or c) use of bistable electro-optical effects with inherent optical memory. The actively addressing approach a) of TN-LCDs will be discussed in Section 6.1.7. In this paragraph the second approach, namely, time multiplexing of supertwisted nematic configurations with steep characteristics will be outlined [30].

During the 1980s several groups discovered different supertwisted nematic configurations which all led to remarkably steep characteristics [5–10, 31]. Common to all of them are nematic helices with twist angles $\varphi > 90°$, that is, supertwisted (STN) LCDs. Figure 6.8a shows the off-state configuration of an STN-LCD with $\varphi = 180°$ twist; whereas Fig. 6.8b depicts its optically uniaxial on-state at voltages far above threshold. Chiral dopants are added to the nematic liquid crystal such that the correct helix sense (left- or righthanded) as well as the approximate desired helical pitch $p \simeq 360° \, d/\varphi$ result. The precise twist angle in STN-LCDs follows from locking the nematic helix to the inner wall aligning layers of the display (Fig. 6.8a).

Figure 6.9 schematically shows typical geometries of three STN-LCD configurations (Table 6.1). They differ with respect to twist angle φ, boundary tilt angle ξ, alignment of entrance and exit polarizers, P_1, P_2 with respect to their adjacent preferred wall aligning directions $\hat{n}_1$, $\hat{n}_2$ and optical retardation $\delta = \Delta n d$ of the nematic LC layer.

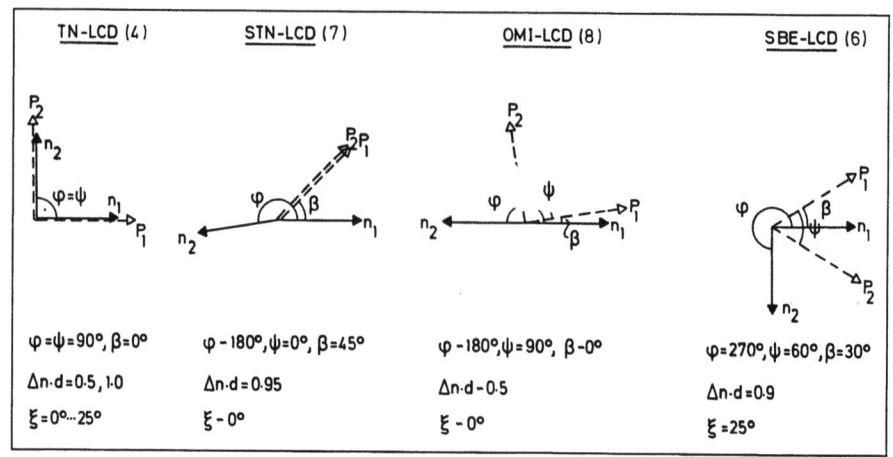

Fig. 6.9. Geometries of twisted nematic (TN) and supertwisted nematic (STN, OMI and SBE) LCDs. n_1, n_2 = boundary aligning directions on the two substrates, P_1, P_2 linear polarizing directions, φ = twist angle, ξ = bias tilt angle.

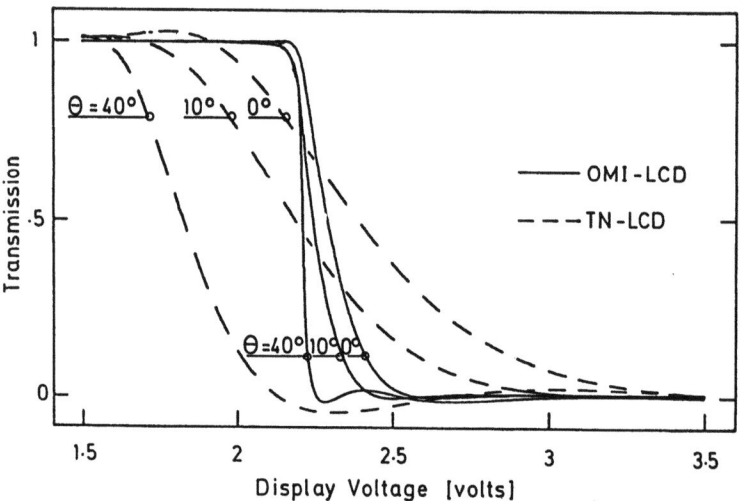

Fig. 6.10. Transmission-voltage characteristics of a TN-LCD and an OMI-LCD for different angles θ of incident light. For vertical light incidence $\theta = 0°$.

To compare the effect of supertwisting a twisted nematic configuration, Fig. 6.10 shows the transmission-voltage characteristics of an optical mode interference (OMI) LCD [8] and of a TN-LCD [4] at different angle θ of light incidence. Both LCDs comprise the same liquid crystal mixture. Instead of the OMI-LCD another

208

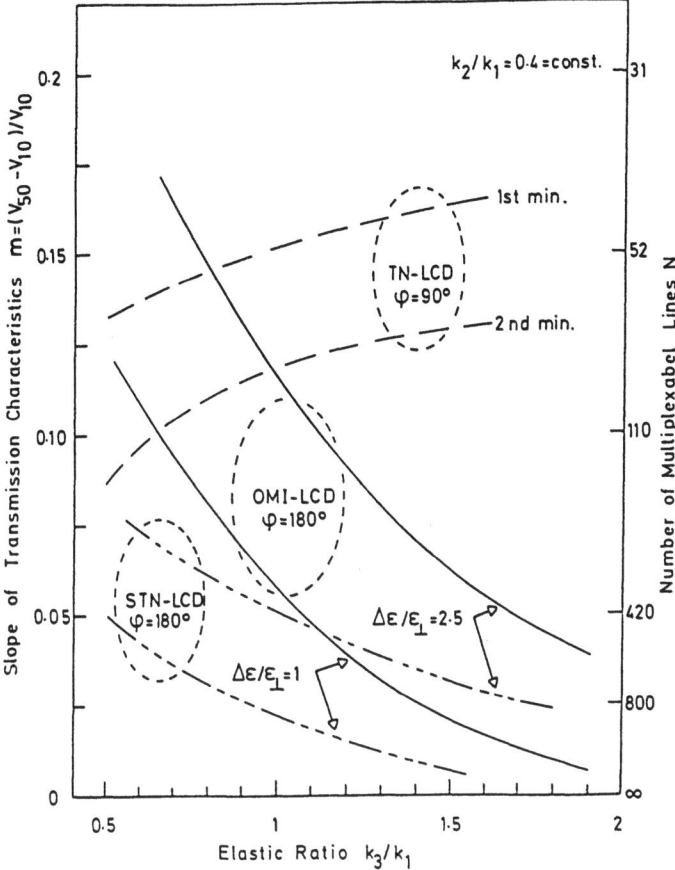

Fig. 6.11. Dependence of the slope parameter m and the multiplexibility N of twisted nematic (TN) and supertwisted nematic (STN and OMI) LCDs on the liquid crystal material parameters k_3/k_1 and $\Delta\varepsilon/\varepsilon_\perp$ for $k_2/k_1 = 0.4 =$ constant.

supertwisted nematic LCD configuration could have been chosen for comparison; the slopes would have been similar. Fig. 6.10 shows i) the much steeper transmission voltage characteristics and ii) the strongly reduced angular dependence of the OMI characteristics compared with the TN-LCD. Thus, the multiplexibility of supertwisted nematic configurations is indeed inherently much better than that of TN-LCDs. However, Fig. 6.10 also illustrates the strongly reduced gray scale capability of supertwist configurations which results from the steep slopes.

Like with TN-LCDs optimal electro-optical performance of supertwisted configurations requires LC materials whose properties are tuned to the specific electro-optical effect and its application. Figure 6.11 shows the influence of the elastic and dielectric material properties of liquid crystals on the slope of the transmission-voltage characteristics of two supertwist configurations with identical twist angles $\varphi = 180°$ (OMI- and STN-LCDs) and of 90° twisted TN-LCDs operated in their first and second minimum [28]. The correlations depicted in Fig. 6.11 were calculated for vertical light incidence and for a constant twist/splay elastic constant ratio $k_2/k_1 = 0.4$ [28]. They confirm the experimental results of Fig. 6.10 and show that both the STN and the OMI supertwist configuration lead to a strong reduction of the slope parameters m. Increasing the twist further leads to a further reduction of m, but also to a slower response and, eventually, at infinitely steep slopes, to fingerprint dislocations in the LCD [32]. Thus, supertwist configurations increase the number of multiplexible lines by about one order of magnitude compared with TN-LCDs. State of the art are 400 multiplexed lines (Fig. 6.2). Prerequisites to achieving this performance are LC materials with large elastic constant ratios k_3/k_1 and/or small dielectric ratios $\Delta\varepsilon/\varepsilon_\perp$. However, the combination of the two, namely, low values of $\Delta\varepsilon$ and high multiplexing rates leads to an increase of the operating voltage which may exceed the capacity of CMOS drivers. This follows from the increase of the select voltage V_s with increasing multiplexing rate (Eq. 6.7) and from the increase of the mechanical [20] and the optical threshold voltage V_t of supertwist configurations which we have recently verified experimentally to be identical [33]:

$$V_t = \pi\{[k_1 + (k_3 - 2k_2)(\varphi/\pi)^2]/\varepsilon_0\Delta\varepsilon + 2k_2(\varphi/\pi)(2d/p)\}^{1/2} \qquad 6.8$$

For nematic liquid crystals which do not comprise chiral additives the natural helical pitch p of the LC material is $p = \infty$. Therefore, and because $\varphi = 90°$, Eq. 6.8 reduces to Eq. 6.3 for TN-LCDs; d = cell gap. From Fig. 6.11 it follows that the way out of this dilemma are liquid crystals with large elastic constant ratios k_3/k_1 (the opposite is required for TN-LCDs, Eq. 6.5 and Fig. 6.11) and large $\Delta\varepsilon$. Compounds which can be designed to ideally combine both properties are polar alkenyls with their double bond in odd side-chain positions [28].

Apart from depending on adequate liquid crystals the multiplexibility of STN-LCDs is also affected by their operating mode. Thus, steep transmission-voltage characteristics in STN-LCDs can also be achieved by operating them in a reflective, single polarizer mode [33].

6.1.6 The Color Problem of Supertwisted LCDs

Both SBE [6] and STN [7] supertwist configurations (Table 6.1 see p. 198, Fig. 6.9) strongly depend on the electrically tunable, large optical retardation $\delta = \Delta nd \simeq 0.90$ of their LC layers. Therefore, both exhibit strong interference colors in either off- (positive contrast) or on-state. Consequently, black-white images as well as full-color images cannot be displayed per se. A first solution of the color problem in supertwist configurations was found with the OMI-LCD [8, 34] (Table. 6.1) which operates with low optical retardation $\delta \sim 0.5$ and crossed linear polarizers (Fig. 6.9). Except for their reduced offstate brightness, the optical appearance of OMI-LCDs is similar to the black-white appearance of TN-LCDs. Moreover, the OMI-effect is much

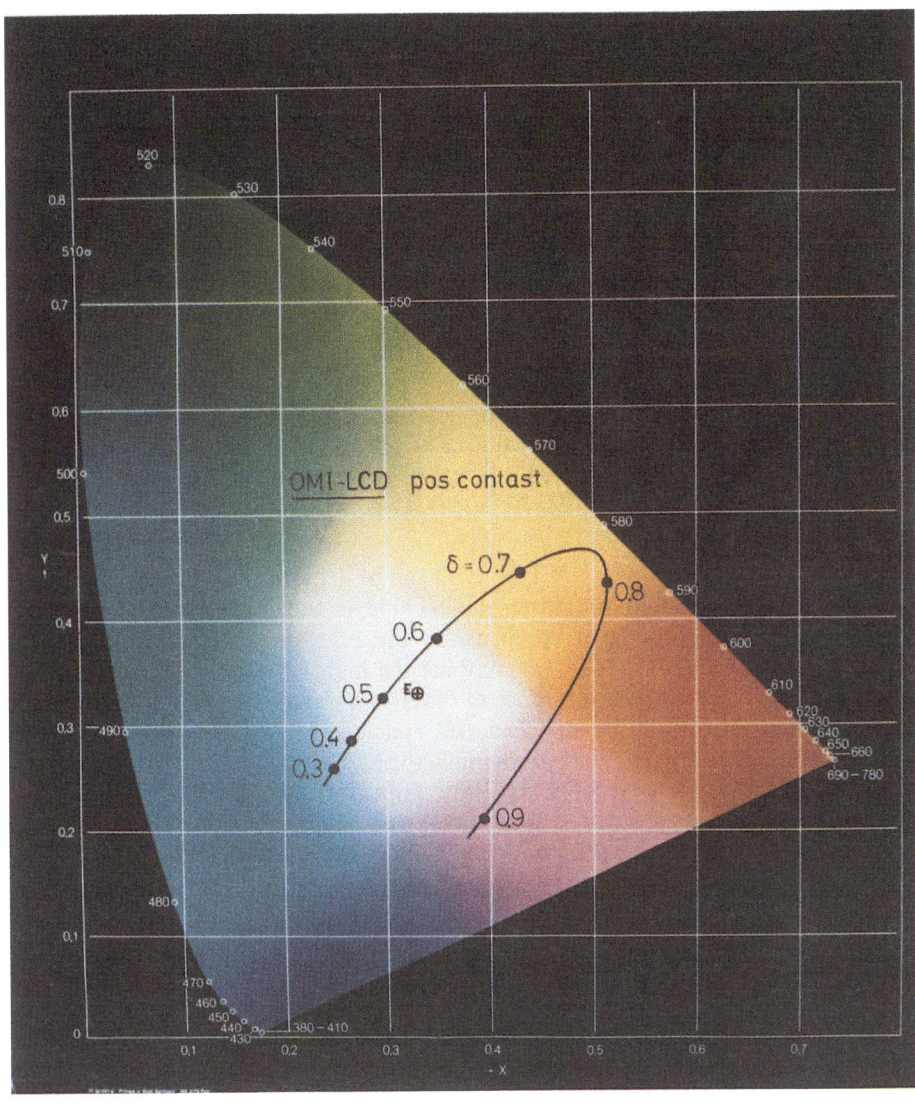

Fig. 6.12. Dependence of the off-state color coordinates of single layer supertwisted nematic (OMI, STN, SBE)-LCDs on the optical retardation $\delta = \Delta nd$ of their LC-layers.

less sensitive to cell gap and temperature variations than the birefringent SBE- and STN-effects [8]. The dependence of the off-state color of positive contrast SBE-, STN- and OMI-LCDs on their optical retardation $\delta = \Delta n \cdot d$ is shown in Fig. 6.12 [8]. The respective on-states are dark blue and black.

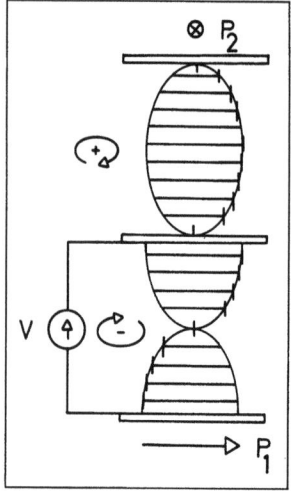

Fig. 6.13. Configuration of a $\varphi = 180°$, double-layer, supertwisted nematic (DSTN) LCD. The helix sense of the addressable STN-LCD (bottom) is opposite to that of the (upper) compensating STN layer.

A black-white, double-layer (DSTN) LCD can be achieved [9] by sandwiching an STN-LCD with a second, non-addressable STN layer such that the optical retardation of the passive STN layer compensates the off-state color of the addressed STN-LCD (Fig. 6.13). DSTN-LCDs combine the high off-state brightness of STN-LCDs with the black-white appearance of OMI-LCDs. However, the three optically flat glass substrates and the matched cell spacings required to achieve optical uniformity in DSTN-LCDs increase manufacturing costs as well as parallax errors.

Instead of using a second STN layer to compensate the off-state color of STN-LCDs, a polymer sheet which acts as an optical retarder can be attached to an STN-LCD [10,35]. This combination, which is easier to manufacture than DSTN-LCDs, has become known as black-white (BW) STN-LCD (Table. 6.1). Because of the temperature dependence Δn (T) of the liquid crystal birefringence (Chapter 2) the optical retardation and therefore the off-state color of STN-LCDs change with temperature. Therefore, unlike in DSTN-LCDs where the temperature dependencies of the two STN layers are matched, or in OMI-LCDs whose temperature dependence is inherently small, BW-STN-LCDs exhibit a residual temperature dependence of their optical appearance.

Since color reproduction strongly increases the information content of a display, a prerequisite to achieving the consequent high multiplexing rates of time multiplexed color STN-LCDs are very steep electro-optical characteristics. From the above it follows that this can be achieved with modern STN LC-materials and twist angles $220° \lesssim \varphi \lesssim 270°$. However, when it comes to displaying color images at video rates, short response times $\tau \lesssim 40$ ms under multiplexed conditions are required, too. That is, response times which are comparably short as the TV frame time $\tau \lesssim 40$ ms. To reach such short response times with highly multiplexed STN-LCDs is a complex task which requires i) liquid crystals with extremely low viscosities and ii) multiplexing addressing techniques which do not generate optical flicker due to the non-rms response of fast STN-LCDs caused by large select voltage pulses. Recently,

Scheffer and Clifton presented an addressing scheme which eliminates row select pulses and the consequent flicker that results from standard addressing techniques [36]. Thus, the possibility exists that OMI-LCDs and later also STN-LCDs will become TV compatible within the near future, provided that proper LC-materials will be developed. However, whether the optical quality of STN-LCDs will reach the very high standard of actively addressed TN-LCDs is questionable.

6.1.7 Full-Color, Actively Addressed TN-LCDs for High Information Content and Fast Response

As shown above, large multiplexing rates, which are prerequisite to realizing high information content supertwist LCDs, are contradictory to the generation of grey levels in STN-LCDs when using conventional addressing techniques. Therefore, and because the response times of high information content STN-LCDs are, so far, not fast enough to reproduce high quality TV color images, alternatives have to be used.

TN-LCDs exhibit a very large contrast between off- and on-state and they appear black-white. Moreover, a large number of grey levels can be reproduced and the response times of directly addressed TN-LCDs comprising low viscous, modern LC materials are fast enough to display TV images. Because of their black-white appearance and their grey level capability TN-LCDs can be combined with color filters such that the full range of colors of the visible spectrum is displayed. This is achieved by mixing the three basic colors red, green, and blue via adjacent TN-LCD pixels. By driving each pixel via electronic switches which are integrated on one of the substrates of the TN-LCD, a matrix array of multiplexible electronic drivers results. Each driver directly addresses a single pixel. Provided very highly resistive LC materials can be achieved, this active addressing scheme combines the favorable direct-drive transmission-voltage characteristics of TN-LCDs (Section 6.1.2) with virtually unlimited multiplexibility of the integrated TN driver matrix. Semiconductor devices for active addressing are metal-insulator-metal (MIM) diodes [37] or thin-film transistors (TFTs) [38, 39] made mainly of amorphous silicon. For a review of active addressing cf. the article of F.C. Luo in [30].

Figure 6.14a shows a cross-section through an actively addressed pixel of a TN-LCD. Each of the three color sections of the pixel are individually addressable by a TFT. The typical area of a single pixel for a direct-view TFT-TN-LCD is 150×50 µm; cell gap $d \cong 5$ µm. Figure 6.14b shows a microphotograph of an active substrate with the three color filters per pixel. Also visible are the (non-transparent) E-shaped TFTs as well as the data- and the select-bus lines to address the TFTs (covered by black masks).

Figure 6.15 shows the equivalent circuit of a TN-LCD pixel addressed via the source S of a thin-film transistor. The liquid crystal layer is simulated by the parallel capacitor C_{LC} and resistor R_{LC}. For a given pixel geometry C_{LC} is determined by $\varepsilon_\perp$ of the LC material (in the off-state), whereas R_{LC} is determined by the residual (ionic) conductivity of the liquid crystal. In a television TFT-TN-LCD each pixel is sequentially addressed with gate pulses such that the capacitance C_{LC} is charged to the

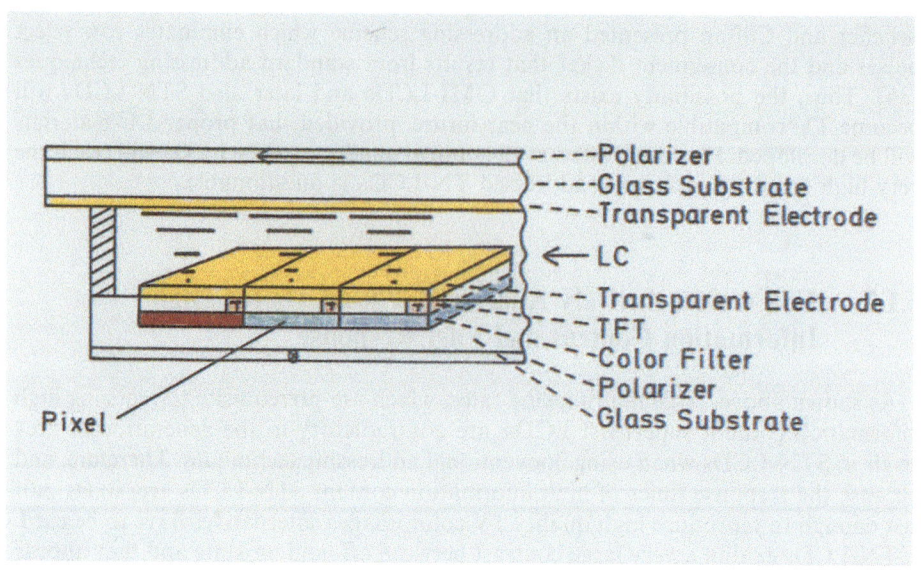

Fig. 6.14. a) Cross-section through an actively addressed picture element of a TN-LCD; TFT = thin film transistor. b) Microphotograph of a TFT substrate with integrated red, green, and blue color filters.

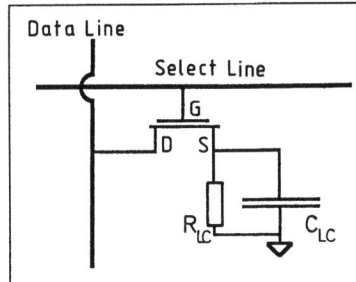

Fig. 6.15. Equivalent circuit of an LCD pixel consisting of parallel C_{LC}, R_{LC} which is actively addressed via the source S of a thin film transistor; D = drain, G = gate.

voltage which corresponds to the desired optical signal (Fig. 6.4). This charge has to be stored by the capacitance C_{LC} of the pixel until the image is either refreshed or changed by a new gate signal. The frequency of change is determined by the frame time $\tau \simeq 40$ ms.

Crucial for the operability of active addressing are liquid crystals with extremely low residual ionic impurities. Therefore, TFT-compatible liquid crystals must be purified to such an extent that specific resistivities $\rho \gtrsim 10^{12}$ Ω m result. Moreover, ρ may not decrease during the lifetime of the LCD due to ionic contamination of the LC material from the LCD substrate surfaces. Low values of R_{LC} would discharge C_{LC} in Fig. 6.15 and, as a consequence, the optical signal would change during the frame time τ. Prerequisites for stable, large resistivities are therefore not only highly resistive LCs with poor ionic solubility, but also LCD substrates which are virtually free of ions. Modern, halogenated TFT-LC materials [40–42] combine the required large ρ-values with low solubility for residual ions from the LCD substrate. This, despite their rather large dielectric anisotropies which are a prerequisite to achieve threshold voltages around 2 volts [32], with a trend towards still lower voltages.

6.2 Field Effects in Chiral Ferroelectric S_C^* Liquid Crystals

Since the discovery of ferroelectricity in chiral smectic C (S_C^*) liquid crystals in 1975 by Meyer et al. [15, 43], great efforts have been made to find electro-optical effects and suitable FLCs which render S_C^* materials applicable in LCDs. These efforts were intensified when Clark and Lagerwall suggested in 1980 to make use of the optical bistability which occurs in long-pitch, surface stabilized ferroelectric liquid crystals (SSFLCs) [16]. Since an unlimited number of lines can be multiplexed with an optically bistable LCD, SSF-LCDs attracted much interest. However, because of a number of ongoing problems, such as shock sensitivity of the LCD alignment, chevron formation perpendicular to the LCD substrates that reduces the off-state brightness of SSF-LCDs, temperature dependence of contrast and difficulties to achieve gray scale, a breakthrough has not yet been made in commercializing ferroelectric LCDs.

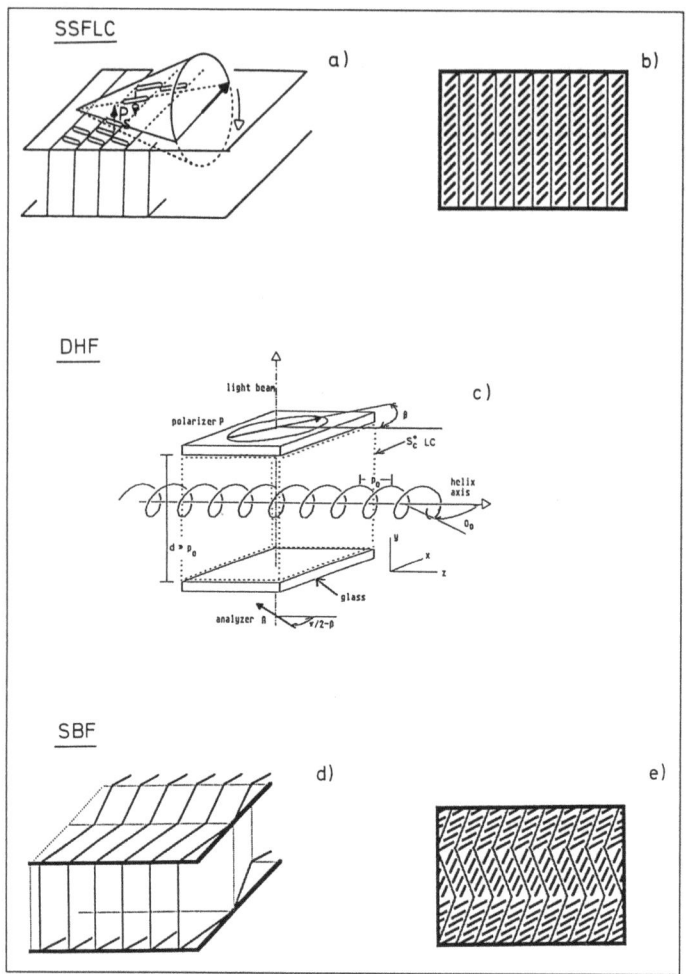

Fig. 6.16. a) Ideal bookshelf geometry of a surface stabilized ferroelectric liquid crystal (SSFLC) layer; b) Director orientation in the substrate plan of an SSF-LCD in one of its two bistable states; c) Deformed helical ferroelectric (DHF) LCD; the helically twisted S_C^* planes are aligned perpendicular to the LCD substrates; d) Likely bookshelf geometry of the short pitch ferroelectric (SBF) configuration; e) Director orientation in the substrate plan of an SBF-LCD in one of its two bistable states.

Figure 6.16a shows the (ideal) bookshelf geometry of SSFLC layers sandwiched between the LCD substrates as suggested by Clark and Lagerwall [16]. The two stable orientations of the director field in the substrate plane result from switching the director on a cone which is determined by the angle 2θ; where θ is the tilt angle of the long molecular axes in the S_C plane (Chapter 1). Switching occurs upon changing the polarity of the applied field. This causes the spontaneous polarization

P_s of the ferroelectric LC layer to reverse direction. Because of the spontaneous polarization of S_C^* materials, short response times in the 50 µs range result for SSF-LCDs at room temperature; i.e., 200 times shorter than the response of TN-LCDs. Figure 6.16b shows one of the two stable director orientations of SSFLCs in the LCD substrate plane. The optical signal which the SSFLC effect generates between crossed linear polarizers is due to the change of the optical retardation upon switching the director between the two stable states. For a review of ferroelectric liquid crystals [44,45].

Because of the complexity of ferroelectric liquid crystals the prospects of finding other electro-optical effects which make use of the interesting properties of S_C^* phases and which may contribute to solve some of the above problems in SSF-LCDs seem good. Thus, Chandani et al. [46,47] have recently discovered a tristable ferroelectric/antiferroelectric electro-optical effect in S_C^* phases with a potential for display applications.

Although optical bistable SSFLCs allow large multiplexing rates, they are – due to their bistability – difficult to address, such that grey levels can be reproduced [48]. Recently, Beresnev et al. [49] have discovered a linear electro-optical effect in ferroelectric LCs. The effect is based on the field-induced deformation of a helical ferroelectric (DHF) structure in short pitch S_C^* liquid crystal layers. Figure 6.16c schematically shows a DHF-LCD [49]. As a consequence of the helix deformation upon applying a voltage to the DHF-LCD, the index ellipsoid of the S_C^* layer rotates in the xz-plane and the LCD transmission changes linearly with increasing voltage. Because of the ideally large tilt angle $\theta \cong 45°$ of DHF-LCDs, their off-state brightness almost reaches the desired 100% at contrast ratios $> 50:1$ [50]. Figure 6.17 shows the linear transmission-voltage characteristics of a DHF-LCD [5]. Moreover, it shows the LC-material-dependent hysteresis of the DHF effect, as well as the very low operating voltages (< 1.8 V) required to drive DHF-LCDs. By actively addressing DHF-LCDs, large multiplexing rates and very short response times in the 10 µs range can be combined with the excellent gray scale capability of DHF-LCDs [50]. The short response times are due to the large P_s-values that can be designed into DHF-LC materials without adversely affecting the DHF configuration. The combination of high multiplexibility, very fast response, gray scale capability, and wide angle of view render actively addressed DHF-LCDs an interesting alternative to actively addressed TN-LCDs for wide angle of view video applications.

Apart from the optically bistable SSF-LCD of Clark and Lagerwall [16] and the tristable effect by Chandani et al. [47], Fünfschilling and Schadt recently presented another optically bistable ferroelectric S_C^* configuration [51]. Prerequisites for their short-pitch bistable ferroelectric (SBF) effect are ferroelectric S_C^* materials which exhibit, unlike SSF materials, a large spontaneous polarization $P_s \sim 60 \, nC/cm^2$ and a short pitch $p \sim 0.4 \, \mu m$ [51]. The molecular configuration of SBF-LCDs is depicted in Fig. 6.16d; whereas Fig. 6.16e shows one of the two stable director configurations in the plane of the SBF-LCD substrate. From Fig. 16d it follows that the S_C^* bookshelfs of the SBF configuration form a slight chevron pattern in the LCD-plane which suppresses helix formation.

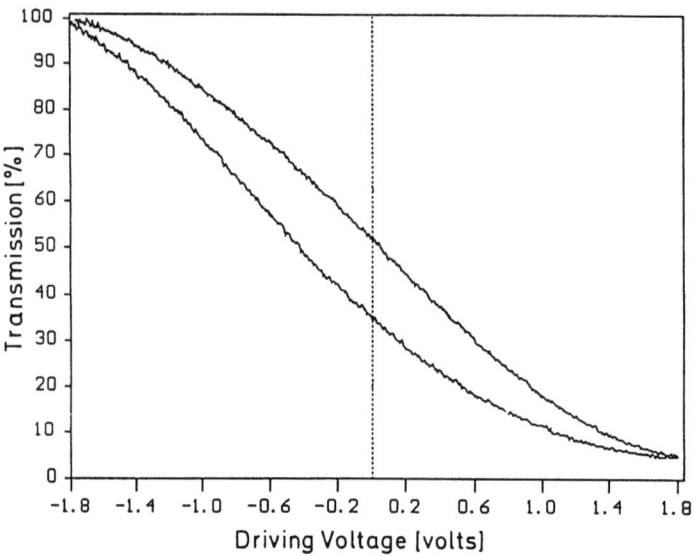

Fig. 6.17. Transmission-voltage characteristics for increasing and decreasing voltage of a DHF-LCD; cell gap $d = 2$ µm; linearly polarized input light intensity = 100%. DHF mixture 6304 from F. Hoffmann-La Roche.

Figure 6.18b shows the dynamics of the transmission of an SBF-LCD upon application of a 20 µs voltage pulse (Fig. 6.18a). Switching between non-transmissive stable state 1 and virtually fully transmissive stable state 2 is achieved within only 10 µs (Fig. 6.18b). From the virtually time-independent transmission of both on- and off-state, it follows that the switching tilt angle of SBF-LCDs is almost as large as the memory tilt angle. Moreover, from the high transmission $T > 85\%$ of state 2 in Fig. 6.18b, it follows that the tilt angle θ in SBF-LCDs reaches, like in DHF-LCDs and unlike in SSF-LCDs, the almost ideal value of $\theta = 45°$. Moreover, θ only weakly depends on temperature, which leads to a desirably weak temperature dependence of display contrast [51]. Recently, it has been shown that SBF-LCDs exhibit contrast ratios up to 80:1, and that stable grey levels can be achieved as well [52].

Patel [53] has recently shown that twisted wave-guiding configurations are not restricted to the twisted nematic effect [4] but can be achieved with S_C^* liquid crystals too. By sandwiching an S_C^* liquid crystal with a tilt angle $\theta \simeq 45°$ between cell surfaces whose surfaces are brushed at an angle of 90° with respect to each other, a 90° twisted smectic C* layer results which exhibits wave-guiding. Unwinding of the smectic helix saturates at about 2 V [53]. Unlike the transmission-voltage characteristics of TN-LCDs, which combine a well defined optical threshold with gray scale (Fig. 6.4), the S_C^*-twisted configuration only exhibits gray scale [53]. In this respect it is similar to the above DHF-effect. However, because of its wave-guiding properties, a stronger

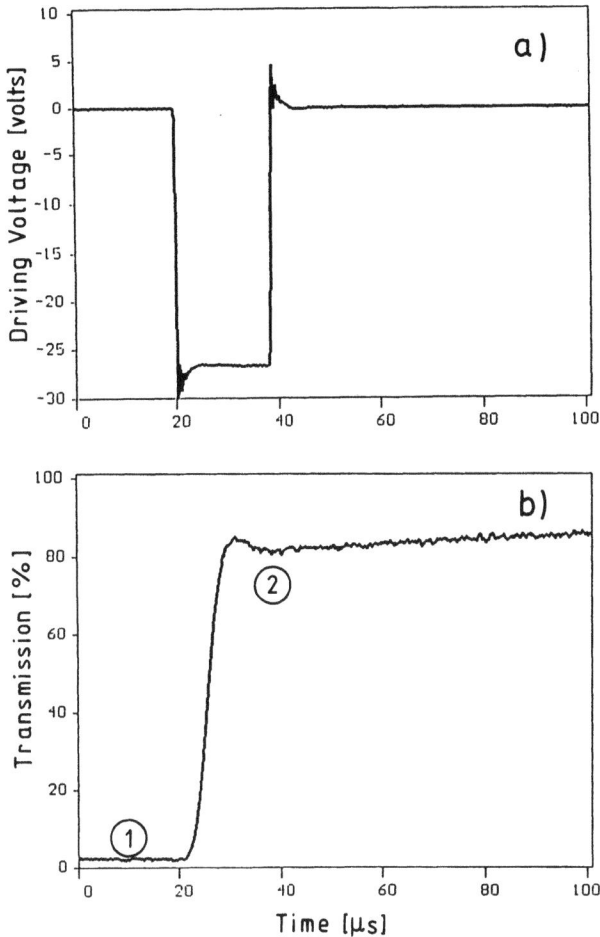

Fig. 6.18. Bistable optical response at 22°C of an SBF-LCD (b) upon application of a 20 μs voltage pulse (a). Cell gap $d = 1.9$ μm; SBF mixture 6430 from F. Hoffmann-La Roche.

angular dependence of its electro-optical characteristics are expected compared with DHF-LCDs.

6.3 Recent Developments in LCD Projection Systems

Because of their flat design and high information content, TFT-TN-LCDs are ideally suited as light modulators in projection systems. By combining them with dichroic mirrors and beam splitters, compact color television LCD projectors result.

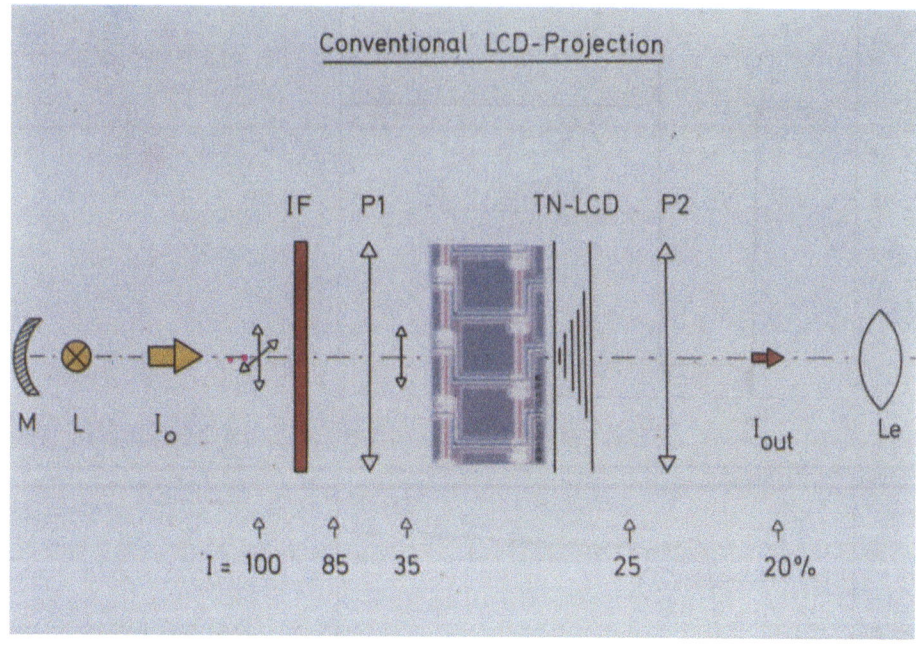

Fig. 6.19. Schematic of a conventional LCD projector for a single color. M = reflector; L = lamp; IF = interference filter (dichroic mirror); P_1, P_2 = linear polarizers attached to TFT-TN-LCD; Le = projection lens.

For a review of conventional LCD projection reference is made to the article of Shields and Bleha [30]. Figure 6.19 shows the principle of a conventional LCD projector comprising a TFT-TN-LCD. Color projection requires splitting of the incident white light into red, green, blue beams, each comprising an LCD modulator. Because of the inherent light absorption of each optical element the output intensity of the system is, at present, much less than 20% (typically 3%) for an unpolarized input intensity $I_o = 100\%$ (Fig. 6.19). A major loss is due to the conversion of unpolarized into linearly polarized light by the (high quality) input polarizer P_1 which absorbs 60%. Another loss factor which strongly depends on the pixel size of the TN-LCD is the geometry of the TFT substrate. For constant display size the amount of light absorbed by bus lines and TFTs increases at the expense of the transmissive pixel area with increasing number of pixels, i.e., with decreasing pixel area (Fig. 6.14b).

An elegant solution which avoids the decrease of brightness of the projector due to TFT- and bus line absorption in the LCD is the introduction of planar microlens arrays by Sharp at the input side of the LCD [54]. By focusing the incident light with microlens arrays onto the transmissive areas of the LCD pixel matrix, such losses can be drastically reduced.

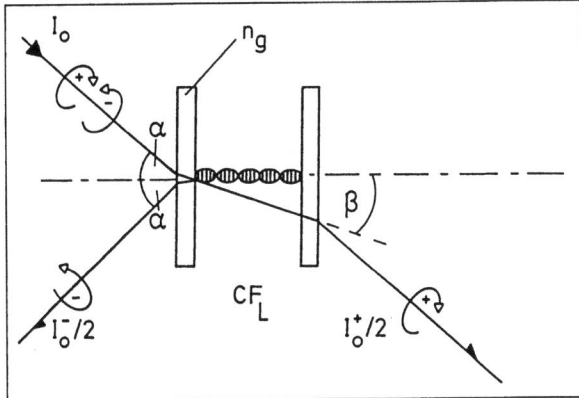

Fig. 6.20. Selective reflection and transmission of unpolarized, incoherent light I_0 incident on a lefthanded, planar aligned cholesteric (filter) layer CF_L.

Recently, we have presented a wide-aperture, color projection principle whose functional parts, namely, polarizers, bandpass filters, and electro-optical modulators consist entirely of liquid crystal elements [17,55,56]. The novel liquid crystal polarized color projection (LC-PCP) concept i) efficiently converts incoherent, unpolarized input light into circularly or linearly polarized light and combines ii) the polarizing properties of planar cholesteric liquid crystal layers with the wavelength selectivity of interference filters. Moreover, since all functional parts of LC-PCPs are non-absorptive, thermal degradation of the optics is strongly reduced. In the context of LC-PCPs it has been shown that TN-LCDs and STN-LCDs, which have so far been operated with linearly polarized light, can also be operated with circularly polarized light. This often simplifies the optics and can improve the electro-optical characteristics of the LCDs [17,33].

Figure 6.20 schematically shows the basic optical properties of planar cholesteric LC layers [57] (Section 6.4.1). Within the wavelength range $\Delta\lambda$, where Bragg reflection at the cholesteric layers occurs, incident, incoherent, and unpolarized light of intensity $I_0 = (I_0^-/2) + (I_0^+/2)$ is partly reflected and partly transmitted by the lefthanded cholesteric layer CF_L. $I_0^-/2$ and $I_0^+/2$ are the left- and right-handed circularly polarized intensities into which I_0 can be decomposed. I_0^+ is transmitted through a left-handed cholesteric layer, whereas I_0^- is totally reflected. The opposite holds for a righthanded layer. The center wavelength of Bragg reflection at vertical light incidence is

$$\lambda_o = p(n_o + n_e)/2 = \bar{n}p, \qquad 6.9$$

where p = cholesteric pitch and n_o, n_e are the refractive indices of the cholesteric liquid crystal. The wavelength range over which selective reflection (transmission) occurs follows from

$$\Delta\lambda = \lambda_o \Delta n / \bar{n} \qquad 6.10$$

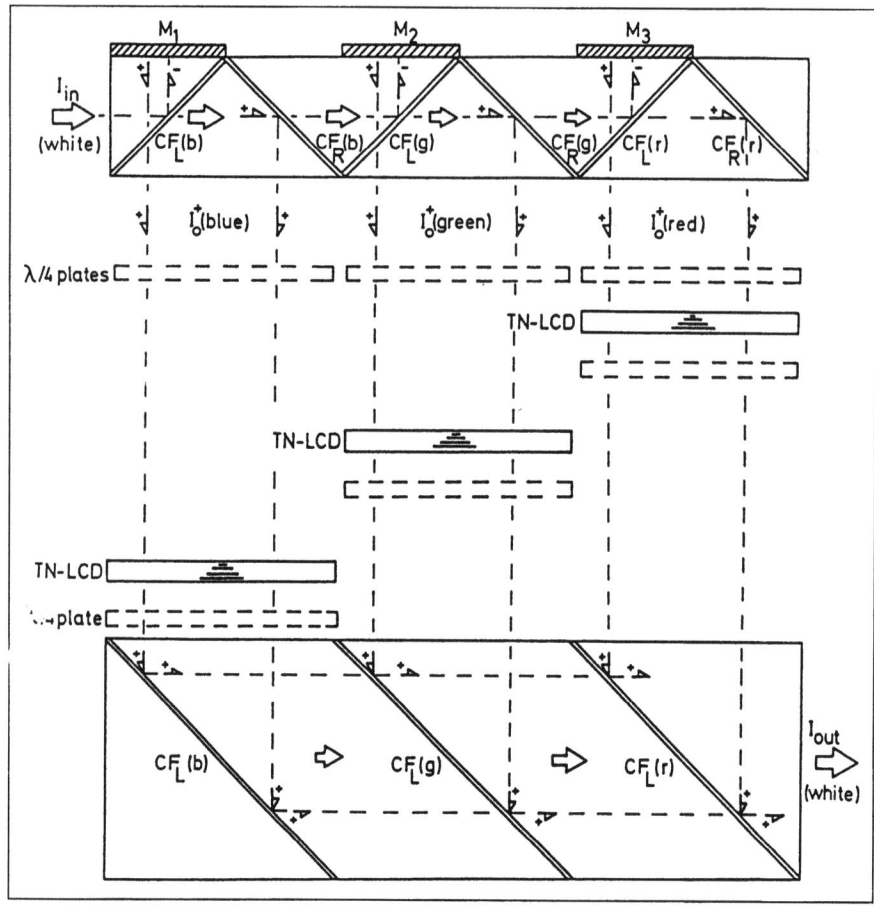

Fig. 6.21. Liquid crystal, polarized color projection (LC-PCP) principle. CF_L and CF_R = left- and righthanded cholesteric filters; M_i = metallic mirrors.

Figure 6.21 shows one among many possible configurations of the liquid crystal polarized color projection (LC-PCP) principle [17,56]. It consists of an input polarizer/filter combination, TN-LCD modulators for the three colors, and an analyzer/beam assembler unit. Ideally, the LC-PCP converts 100% of unpolarized, white input light $I_{in} = (I_{in}^-/2) + (I_{in}^+/2)$ into the three monochromatic, polarized beams I_o^+ (blue), I_o^+ (green), I_o^+ (red) which are in Fig. 6.21 recombined by the LC-analyzer into the single output beam I_{out}. The polarizer/filter combination at the input of the LC-PCP in Fig. 6.21 consists of three matched pairs of left- and righthanded cholesteric notch filters ($CF_L + CF_R$) which are tuned to the respective center wavelengths $\lambda_o = 470$ nm, 530 nm and 610 nm and which comprise temperature independent cholesteric LCs [17]. The LC-PCP operates as follows: within its (blue) selective reflection band the first notch filter combination $CF_L(b) + CF_R(b)$

reflects the respective left- and righthanded circularly polarized fractions of unpolarized blue input light $I_{in}(b) = (I_{in}^+(b)/2) + (I_{in}^-(b)/2)$ in two different directions. The totally reflected intensity $I_o^+(b)$ depends on the bandwidths of the filters (Eq. 6.10). Upon reflection at the metallic mirror M1, the handedness of (b) changes from left to right and can thus pass the first filter CF_L (b). The righthanded input fraction $I_o^+(b)$, which is transmitted by $CF_L(b)$, is reflected by the second filter $CF_R(b)$ (Fig. 6.21). Unpolarized input light with wavelengths outside of the selective reflection band of the blue notch filter is transmitted into the next two polarizing/filter stages of the LC-PCP. Thus, within the selective reflection bands of its filters, the input optics of the LC-PCP in Fig. 6.21 selectively converts the unpolarized input intensity $I_{in} = I_{in}(b) + I_{in}(g) + I_{in}(r)$ fully into the righthanded, circularly polarized, monochromatic, blue, green, and red intensities $I_o^+(b)$, $I_o^+(g)$ and $I_o^+(r)$. With additional $\lambda/4$-plates the circularly polarized outputs can be converted into linearly polarized light (Fig. 6.21).

The analyzer/beam assembler of the LC-PCP in Fig. 6.21 is also made of a combination of three lefthanded cholesteric filters. They selectively reflect and assemble $I_o^+(b)$, $I_o^+(g)$ and $I_o^+(r)$ into the single output beam I_{out}. Since the helical sense of I_o in Fig. 6.21 is determined by the sequence of the handedness of the notch filters in the input stage of the LC-PCP, the handedness of the CFs in the analyzer has to be selected accordingly.

When neglecting reflection losses at the optical interfaces as well as losses due to scattering at dislocations in the cholesteric filters, the conversion efficiency of the liquid crystal polarized color projector (LC-PCP) concept ideally reaches $I_{in}/I_{out} = 100\%$ [17,56]. Apart from the LC-PCP optics depicted in Fig. 6.21 other transmissive as well as reflective configurations using the new projection principle are feasible [17,56]. The recent development of cholesteric liquid crystalline sidechain polymers with high glass-transition temperatures [58] will further extend the operating temperature range of cholesteric filters/polarizers from now $0°C...80°C$ to $-30°C...100°C$. Moreover, by using negative dielectric anisotropic cholesteric LCs, field-induced planar cholesteric filters with ideal, dislocation free optical properties have recently been realized [59]. By combining the LC-PCP concept with microlens arrays which focus the light onto the LCD pixels [54], very bright high information content LCD projectors become feasible.

Conclusion

The rapidly expanding LCD technology has come close to one of its ultimate goals: the flat, low weight, full-color, fast responding, large-area, high information content, affordable display. This goal, which many thought to be too far-fetched when today's nematic field-effect technology dawned in the early 1970s with the discovery of the twisted nematic effect, now begins to materialize. Not only the realization of flat color television LCDs that exceed the present 14" diagonal, direct-view LCD screens, but also bright and compact, large-area projection LCDs with high-definition TV resolution seem feasible within the next few years. Their compactness, low weight, compatibility with low-power and low-voltage CMOS driving circuitry, as well as their high optical image quality will open up new applications for LCDs which could otherwise not, or only marginally be realized. Recent examples are laptop and notebook computers; space-saving and flicker-free, full-color LCD computer terminals with low power consumption; high-resolution graphics displays for navigation systems in avionics and cars; electronically tunable windows, etc., Moreover, the recently discovered generation of planar LC-director aligning pattern by linear photopolymerization of LCD substrates [60] opens up interesting possibilities for realizing new optical and electro-optical devices. Thus, different electro-optical effects can be combined on the same LCD substrate and the generation of stereo images becomes feasible.

References

1. Schadt M (1992) Field effect liquid crystal displays and liquid crystal materials: Key technologies of the 90s. Displays 13, No 1:11–34
2. Fünfschilling J (1991) Liquid crystals and liquid crystal displays. Condensed Matter News 1, No. 1
3. Boller A, Scherrer H, Schadt M, Wild P (1972) Low electro-optical threshold in new liquid crystals. Proc IEEE 60:1002
4. Schadt M, Helfrich W (1971) Voltage dependent optical activity of a twisted nematic liquid crystal. App Phys Lett 18:127
5. Waters CM, Raynes EP, Brimmel V (1983) Highly multiplexible dyed liquid crystal displays. SID Proc Japan Display '83:396
6. Scheffer TJ, Nehring J (1984) A new, highly multiplexible liquid crystal display. Appl Phys Lett 45:1021
7. Kando Y, Nakagomi T, Hasegawa S (1985) German Patent DE 3503259 AI
8. Schadt M, Leenhouts F (1987) Electro-optics of a new, black-white and highly multiplexible LCD. Appl Phys Lett 50:236
9. Katoh K, Endo Y, Akatsuka M, Ohgawara M, Sawada K (1987) Application of retardation compensation: A new highly multiplexible black-white LCD with two supertwisted nematic layers. Jpn J Appl Phys 26:L1784
10. Okumura O, Nagata M, Wada K (1987) ITEJ Tech Report II:27
11. Doane JW, Vaz NA, Wu BG, Zumer S (1986) Field controlled light scattering from nematic microdroplets. Appl Phys Lett 48:269
12. Fergason J (1986) Polymer encapsulated nematic liquid crystals for displays and light control applications. Digest SID '86:88
13. Fujisawa T, Ogawa H, Maruyama K (1989) Electro-optic properties and multiplexibility of PN-LCDs. Proc Jpn Display '89:690
14. Kunigita M, Hirai Y, Ooi Y, Niiyama S, Asakawa T, Masumo K, Kumani H, Yuki M, Gunjima T (1990) A full color projection TV using LC/polymer composite light valves. Digest SID '90:227
15. Meyer RB, Liebert L, Strzelecki L, Keller P (1975) Ferroelectric liquid crystals. J. Phys Paris 36:L69
16. Clark NA, Lagerwall ST (1980) Submicrosecond bistable electro-optic switching in liquid crystals. Appl Phys Lett 36:899
17. Schadt M, Fünfschilling J (1990) New liquid crystal polarized color projection principle. Jpn J Appl Phys 29, No. 10:1974
18. Berreman DW (1974) Optics in stratified and anisotropic media: 4 × 4 matrix formulation. J Opt Soc Am 63:1374

19. Gooch CH, Tarry HA (1975) The optical properties of twisted nematic liquid crystal structures with twist angles $\leq 90°$. J Phys D 8:1575
20. Schadt M, Gerber P (1982) Class specific physical properties of liquid crystals and correlations with molecular structures and electro-optical performance in TN-LCDs. Z Naturforsch 37a:165
21. Meyerhofer D (1977) Optical transmission of liquid crystal field-effect cells. J Appl Phys 48:1179
22. Pohl L, Weber G, Eidenschink R, Baur G, Fehrenbach W (1981) Low Δn twisted nematic cell with improved optical properties. Appl Phys Lett 38:497
23. Jakeman E, Raynes EP (1972) Electro-optical response times in liquid crystals. Phys Lett 39A:69
24. Schadt M, Mueller F (1978) Physical properties of liquid crystal mixtures and performance in twisted nematic LCDs. IEEE Trans Electron Devices ED25:1125
25. Gerber P, Schadt M (1980) Viscous properties of different liquid crystal classes and dynamic performance in TN-LCDs. Z Naturforsch 35a:1036
26. Berreman DW (1975) Liquid crystal twist cell dynamics with backflow. J Appl Phys 46:3746
27. Doorn CZ van (1975) Dynamic behaviour of twisted nematic liquid crystal layers in switched fields. J Appl Phys 46:3738
28. Schadt M, Buchecker R, Müller K (1989) Material properties, structural relations with molecular ensembles and electro-optical performance of new bicyclohexane liquid crystals in LCDs. Liquid Crystals 5:293
29. Alt PM, Pleshko P (1974) Scanning limitations of liquid crystal displays. IEEE Trans Electron Devices ED21:146
30. Scheffer T, Nehring J (1991) Twisted nematic and supertwisted nematic mode. LCDs In: Bahadur B (ed) Liquid Crystals Applications and Uses, vol 1. World Scientific, New York
31. Raynes EP (1986) The theory of supertwist transitions Mol. Cryst Liq Cryst Lett 4:1
32. Moia F, Schadt M (1991) New generation of mixtures for high information content LCDs, such as OMI-, STN-, and TFT-TN-LCDs. Proc SID 32, No 4:361
33. Schadt M (1991) Novel supertwisted nematic LCD operating modes and electro-optics of generally twisted nematic configurations Proc SPIE/SPSE 1455:214
34. Kawazaki K, Yamada K, Mizunoya K (1987) High display performance black-white supertwisted nematic LCD. Digest SID '87:391
35. Takiguchi Y, Kanemoto A, Iimura H, Enomoto T, Iida S, Toyooka T, Ito H, Hara H (1990) Achromatic supertwisted nematic LCD with polymeric liquid crystal compensator. SID Eurodisplay 90:96
36. Scheffer TJ, Clifton B (1992) Active addressing method for high contrast video-rate STN-LCDs. Digest SID 23:228
37. Morozumi S, Ohta T, Araki R, Kubota K, Ono Y, Nakazawa T, Ohara H (1983) A 250×240 element-LCD addressed by lateral MIM: SID Digest Japan Display '83:404
38. Brody TF, Luo FC, Davies DH, Greeneich EW (1974) Operational characteristics of a $6'' \times 6''$ TFT-matrix array LCD. Digest SID '74:166
39. Morozumi S, Oguchi K, Misawa T, Araki R, Oshima H (1984) 4.25" and 1.51" black-white and full color video displays addressed by poly-Si TFTs. Digest SID '84:316
40. Schadt M, Buchecker R, Villiger A (1990) Synergisms, structural-materials relations and display performance of novel, fluorinated alkenyl liquid crystals. Liquid Crystals 7:519
41. Sugimori (1980) German Offenlegungsschrift DE 3102017C2
42. Finkenzeller U, Kurmeier A, Poetsch E (1989) New fluorinated liquid crystaline compounds with positive dielectric amisotropy. Proc Freiburger Arbeitstagung
43. Meyer RB (1977) Ferroelectric liquid crystals, a review. Mol Cryst Liq Cryst 40:33

44. Fukuda A, Ouchi Y, Arai H, Takano H, Ishikawa K, Takezoe H (1989) Complexities in the structure of ferroelectric liquid crystal cells; the chevron structure and twisted structures. Liquid Crystals 5:1055
45. Rieger H, Escher C, Illian G, Jahn H, Kaltbeitzel A, Ohlendorf D, Rösch N, Harada T, Weippert A, Lüder E (1991) FLCD showing high contrast and high luminance. Digest SID '91:396
46. Chandani ADL, Hagiwara T, Suzuki Y, Ouchi Y, Takezoe H, Fukuda A (1988) Tristabel switching in surface stabilized ferroelectric liquid crystals with large spontaneous polarization. Jpn J Appl Phys 27:L729
47. Chandani ADL, Gorecka E, Ouchi Y, Takezoe H, Fukuda A (1989) Antiferroelectric chiral smectic phases responsible for the tristable switching. Jpn J Appl Phys 28:L1265
48. Hartmann WJAM (1989) Charge controlled phenomena in the surface stabilized ferroelectric LC-structure. J Appl Phys 66:1132
49. Beresnev LA, Chigrinov VG, Dergachev DI, Poshidaev EP, Fünfschilling J, Schadt M (1989) Deformed helix ferroelectric liquid crystal display: A new electro-optical mode in ferroelectric C* liquid crystals. Liquid Crystals 5:1171
50. Fünfschilling J. Schadt M (1989) Fast responding and highly multiplexible distorted helix ferroelectric LCDs. J Appl Phys 66:3877
51. Fünfschilling J, Schadt M (1991) New, short-pitch bistable (SBF) liquid crystal displays. Jpn J Appl Phys 30:741
52. Fünfschilling J, Schadt M (1991) Short pitch bistable ferroelectric liquid crystal displays. Digest SID 11th Internat Display Research Conf, San Diego: 183
53. Patel JS (1992) Ferroelectric liquid crystal modulator using twisted smectic structure. Appl Phys Lett 60:280
54. Hamada H, Funada F, Hijikigawa M, Awane K (1992) Brightness enhancement of an LCD projector by a planar microlens array. Digest SID 92:269
55. Belayev SV, Schadt M, Barnik MI, Fünfschilling J, Malimoneko NV, Schmitt K (1990) Large aperture polarized light source and novel liquid crystal display operating modes. Jpn J Appl Phys :L634
56. Fünfschilling J, Schadt M The efficient optics of liquid crystal polarized color projection Optoelectronics Devices and Technologies (OP-DET) 7, Nr. 2, 263 (1992)
57. Belyakov (1992) Diffraction optics of complex structured periodic media. In (ed) Partially Ordered Systems. Springer, New York
58. Häberle H, Leigerber H, Maurer R, Miller A, Stohrer J, Buchecker R, Fünfschilling J, Schadt M (1991) Right- and left circularpolarized colorfilters made from cross-linkable cholestericsilicones. Digest SID 11th Internat Display Research Conf :57
59. Schadt M, Linear and nonlinear liquid crystal materials, electro-optical effects and surface interactions: Their applications in present and future devices. Liquid Crystals 14, 73 (1993).
60. Schadt M, Schmitt K, Kozinkov V, Chigrinow V, Surface-induced parallel alignment of liquid crystals by linearly polymerized photopolymers. Jpn J Appl Phys 31, 2155 (1992).

Subject Index

A
absorption edge 175
aggregates, micellar 146
aggregation 146
alignment
– homeotropic 79
– planar 79
alignment of nematics
– by electric and magnetic fields 70, 177
– by surfaces 78
alkyl spacers 105
amphiphiles
– anionic 147
– cationic 147
– nonionic 147
– mesomorphic 147
amphotropic 4
anchoring energy 83
anisotropic
– phase 117
– solvents 175
anisotropy
– optical 57
– of the molecular polarizability 59, 71
– of the diamagnetic susceptibility 68
– of the dielectric permittivity 71
– of the electric conductivity 93
applications
– LCD devices:
– – TN 195, 206
– – STN 206
– – OMI 208
– of LC polymers 114
– of lyotropic liquid crystals 168

B
backbone, polymer
– orientation 122, 123
– stiffness 109, 110

bend deformation 77, 78, 85, 91
– elastic constant k_{33} 77, 201
biaxial liquid crystals 63
– nematics 65
– smectics 63
biaxiality 63
bilayer structure
– membranes 168
birefringence 57, 58
– negative 58
– of cholesteric phases 62
– of discotic phases 61
– of nematic phases 59
– of smectic A and B phases 62
– positive 58
bistability of ferroelectric smectic C* phases 89, 216
blue phase 12
bond angles 188, 189
bond lengths 188, 189
book-shelf geometry 89, 90, 216
boundary conditions 79
bowlic 44
Bragg reflection of cholesterics 66
Bragg's equation 13, 66
broken fan-shaped texture 19, 24

C
C* phase, smectic 24
calamitic LC side-chain polymers 116
– chemistry of 26
– liquid crystals 5, 6
calorimetric investigations 51, 127
– transition enthalpies 52
– transition entropies 56
calorimetry 51

chain melting 148, 154
chevron domain 98
chiral
– nematics 11
– smectics 24
chiral nematic liquid crystals 11, 65
chiral smectic liquid crystals 24
– smectic C* phase 25, 67
chirality, molecular 67
cholesteric liquid crystals 11
– birefringence 62
– circular polarized light 65
– compensated 180
– dielectric anisotropy 73
– electric field effects 84
– Grandjean plane texture 80
– helical twisting power 67
– optical properties 65
– optical rotation 66
– origin of name of 11
– pitch of 11
– reflection wavelength 65
– selective reflection by 65
cholesteric pitch 11
chromonics 149
circular dichroism of cholesterics 66
clearing point 4
clearing temperature 4, 56
cloud point 153
cmc 146
co-surfactant 154
columnar phases 5, 43
combined main chain/side chain polymers 127
conductivity anisotropy of, in nematics 93
continuum theory 35, 77

227

contrast 200
copolymer, liquid crystalline 123
counter ion 149
critical field for untwisting of the helix
- of cholesteric phases 84
- of smectic C* phases 89
critical micelle concentration 146
cubic mesophases 5, 12, 49

D
Debye length 163
Debye's equation 73
decay times 83, 84, 203, 217
decoupling of polymer main chains and mesogenic groups 120
defects
- in cholesterics 12
- in liquid crystals 7, 9
- in nematics 9
- in smectics 17
deformation
- in liquid crystals 77
- - bend 78, 91
- - splay 78, 91
- - twist 91
deformed helix
- in ferroelectric liquid crystal displays 216, 217
degree of multiplexing 204
- of polarization 179
- of polymerization 122
degree of order
- in nematics 6, 173
- in smectics 39, 54
diamagnetic anisotropy 68, 184
- susceptibility 68, 184
dichroic ratio 179
dielectric properties 70
- anisotropy 71
- constant 71
- loss 74
- permittivity 71
- relaxation 74, 131
- - time 98
- reorientation 81

Differential scanning calorimetry 51, 127
dimesogenic polysiloxanes 123
dipole coupling 185
director 7
director reorientation 131
disc-like molecules 44
- mesogenes 133
disclination 7, 9
discotic nematic phase 42
discotic liquid crystal 5, 42
- chemistry of 44
- LC polymers 133
- molecular statistical theory of 46
- molecular structure 45
- phase structure 42
- types of 42
dispersion forces 39
dispersion interaction 39
displays, liquid crystal 195
Domains, Williams 95
double refraction
- see birefringence
dye, dichroitic 180
dynamic scattering 95
dynamical effects in nematics 83

E
Econol 116
elastic constants 77, 86
- bend 77
- splay 77
- twist 77
elastomers, liquid crystalline 136
electric conductivity 92
- anisotropy of the 93
electric field effects 80, 84
electric instabilities 95
electro-optical effects 197
electrohydrodynamic instabilities 95
- deformations 95
- effects 95
electron microscopy 130
Electron spin resonance spectroscopy 188
enthalpies for mesophase transitions 52

entropies for mesophase transitions 54, 108

F
fan shaped texture 18, 19
ferroelectric coefficients 90
ferroelectric liquid crystal polymer 131, 133
ferroelectric liquid crystals 25, 88, 215
- deformed helix 216
ferroelectric switching in polymers 131
fiber spinning 113
fiber, ultra-high-strength 114
first-order transitions in liquid crystals 53
flexible spacers 107
flexo-electricity 90
flexoelectric effect 90
fluorescence spectroscopy 182
focal conic texture 16
Frank elastic constants 77
Frank elastic energy 77
Fredericksz transitions 131
free energy of elastic deformation 77, 85

G
gel phase 152
glassy state 119, 130
Grandjean texture 80
gray scale 197
guest-host effect 198

H
^{2}H-NMR of liquid crystalline polymers 130
heats of transition 52
helical twist sense 12, 212
- twisting power 67
helix, cholesteric 11
- pitch 11, 65
herringbone structure 22
hexagonal mesophases 17, 21
high-information-content display 213
historical development 1, 143

HLB-parameter 164
homeotropic
  alignment 79, 177
– phase 177
homogeneous
  alignment 79, 177
– phase 177
homologous series 35
hydrodynamic effects 95, 99
– instability 95
hydrophilic group 147
hydrophilic
  interactions 162
hydrophobic group 147
– effect 145
hyper-fine splitting 189

I
index of refraction 58
indicatrix 57, 65
induced polarization 71
infrared spectroscopy 177
– dichroism 177
interactions 162
– dispersion 39, 162
– electrostatic 163
– hydrophobic 163
– solvation 163
– steric 163
interactions, direct dipole-dipole
– intermolecular 185
– intramolecular 185
intermicellar forces 162
intermolecular direct di-
  pole-dipole
  interactions 185
intermolecular forces 36
intramolecular direct di-
  pole-dipole
  interactions 185
ion mobility 93
IR dichroism 177
iridescent colors of
  cholesterics 12, 65

K
Kevlar fiber 114, 115
Kraft point 148
Kraft temperature 148

L
lamellar mesophases 150
Landau theory 54
layer normal 86
layer tilting 14, 17, 21
LCD 195
LCD projection 219
LCD television 213, 223
length-to-breadth ratio 38
liquid crystal 1, 4
liquid crystal displays 195
liquid crystalline
  elastomers 136
liquid crystalline
  polymers 103
liquid crystalline
  solvents 175, 176
long-range orientational
  order 6, 39, 173
long-range positional
  order 13, 39
Lorenz-Lorentz
  equation 58
lyotropic cholesteric
  mesophases 150
lyotropic liquid crystalline
  polymers 108
lyotropic liquid crystals 1, 4, 143
– cubic 151, 152
– gel 152
– hexagonal 150, 151, 158
– lamellar 150, 151, 159
– nematic 150, 151, 157

M
magnetic field effects
– on nematics 70
magnetic properties 68
– anisotropy of the
  susceptibility 68
– diamagnetic
  susceptibility 68
magnetic susceptibility 68
– anisotropy of the 68
Maier-Meier theory 71
Maier-Saupe theory 39
main chain LC
  polymers 104
marbled texture 10, 11
McMillan theory of
  smectics 39

mean field theory 36
mesogen
– disc-like 5
– rigid rod-like 5
mesomorphic phase 4
mesophase 4
– amphiphilic 143
– thermotropic 4
micelle 146
– normal 147
– reverse 147
middle phase 145
miscibility of LC
  polymers 111
molecular axial ratio 110
molecular
  polarizability 58
– anisotropy of 59
– average 58, 71
molecular statistical theory
– of discotic phases 46
– of the nematic state 35
– of the smectic state 39
mosaic texture 22, 23
Mößbauer
  spectroscopy 191
multicritical point 54
multiplexing 204

N
neat phase 145
nematic 5, 150
nematic liquid crystals 6, 120, 151
nematic marbled
  texture 10
nematic phase 6, 120, 151
– birefringence 59
– dielectric anisotropy 71
– elastic properties 77
– electric properties 70, 92
– order parameter 6
– textures 9, 10, 157
nematic Schlieren
  texture 7, 9, 157
nematic thread-like
  texture 10
nuclear magnetic resonance
  spectroscopy 184
NMR spectroscopy 184

O

odd-even-effect 56
oily streak texture 12, 159
OMI-LCD 208
Onsager equation 70
optical activity 66
optical anisotropy 57
optical contrast 200
optical properties of liquid crystals 57
optical properties of twisted liquid crystals
– chiral smectic phases 67
– cholesteric phases 65
optical rotation 66
optical textures 9ff., 157
optically biaxial liquid crystals 63
optically uniaxial liquid crystals 57
– extraordinary ray 58
– ordinary ray 58
– refractive index 58
order parameter
– nematic 6, 173
– smectic 39, 54
orientation polarization 71
orientational order parameter 6, 59
– of LC polymers 122
– of nematics 8, 38, 173, 188

P

packing fraction 37, 38
phase sequence 26
– of lyotropic liquid crystals 150
phase transitions
– first order 52
– second order 53
piezoelectric properties 137
pitch, helical
– of cholesterics 11
– of smectic C* phases 67
pixel 196, 213
planar orientation 79, 177
planar-twisted cell 81, 82
plastic crystals
– cubic 3
– hexagonal 3
Poiseuille's law 99
polar molecule 14, 30
polarizability
– anisotropy of the 58
– average 58
– longitudinal 59
– parallel and perpendicular to the molecular long axis 59, 71
– transversal 59
polarization, spontaneous 89
polarized optical spectra 173, 177, 180
polarizing microscopy 7
Poly-(benzyl-L-glutamate) 111
polygonal texture 18
polymer backbone 122, 124
polymer, liquid crystalline 103
– lyotropic 108
– main chain 104
– side chain 116
polymorphic smectic phases 26
polymorphism
– of smectic phases 26
positional order parameter 39
pressure dependence 52
pretransitional smectic order 101
pyramidic 44

R

reentrant behaviour 26
reentrant nematic phase 26
reflection of light 12, 25, 65, 131
refractive index 57
– anisotropy 58
– extra ordinary 58
– ordinary 58
refractive indices of mesophases 57
Reinitzer, Friedrich 1
relaxation time 74
relaxation, dielectric 74
– in LC polymers 131

response time 83, 89, 203, 217
rheological properties
– of LC polymers 113
– of nematics 99
rigid rod polymers 108
rise time 83, 203
rotational viscosity 83, 87, 203

S

sandinic liquid crystals 5, 47, 135
Schlieren texture 9, 16, 23
second-order transition 53, 54
selective reflection
– of cholesterics 66, 131, 221
– of smectic C* phases 68
sequence rule for polymorphic liquid crystals 26
side chain LC polymers 116
side-on fixed mesogen in LC polymers 126
smectic A phase, birefringence 62
smectic A phases 13
– B phases 17, 19
– C phases 14
– E phases 20
– F phases 17, 20
– G phases 17, 20
– H phases 20, 21
– I phases 17, 20, 21
– J phases 17, 21
– K phases 20, 21
– L phases 17, 21
– M phases 17, 21
smectic B phase, birefringence 62
smectic C, Schlieren texture 16
smectic liquid crystals 5, 12, 21
– chiral 24
smectic order 39
smectic order parameter 39, 54
smectic phases, thermotropic 12

230

– lyotropic 150
spacer concept for LC
  polymers 116
spacer, flexible 118
spectroscopy, LC solvents
  for 175
spin-spin coupling
  constant 185
splay deformation
– of nematic phases 77, 91
– of smectic phases 85
splay elastic constant
  k_{11} 77, 201
spontaneous
  polarization 89
statistical theory
– of nematics 36
– of smectics 39
STN-LCD 206
structure
– of cholesteric phases 11
– of lyotropic phases 150
– of nematic phases 6, 8,
  120, 151
– of smectic phases 12, 128
supertwisted nematic LC
  display 206
surface alignment 78
surface-stabilized
  ferroelectric LC 215
surfactant 79, 146, 150
susceptibility, magnetic 68
switching in electric
  fields 83, 87, 89, 203,
  218
– of LC polymers 131
switching time 83, 87, 89,
  133, 203, 218

T
texture
– fan-shaped 18, 19
– focal conic 16
– Grandjean 80
– oily streaks 12, 159
– Schlieren 9, 16, 23
TFT-TN-LCD 213
thermotropic liquid
  crystals 4
thin-film transistor 213,
  214
threaded texture 10
threshold field 82, 86, 96
threshold voltage 82, 96
tilt angle for smectic
  C phases 14
tilted smectic phases 14,
  17, 21
time multiplexing of
  LCDs 204
TN-LCD 195, 200
transition moment 179
translational order
  parameter, see positional
  order parameter
tricritical point 54
twin-dimers 107
twist
– in cholesterics 11, 65
– in smectic C* phases 24,
  67
twist deformation 77
twist elastic constant
  k_{22} 77, 201
twisted nematic liquid
  crystal display 195, 197
twisting power, helical 67

U
ultraviolet
  spectroscopy 180
– nematic solvents for 175
uniaxial liquid crystalline
  phases 57
– cholesteric 62
– nematic 59
– smectic 62
untwisting of the helix
– of cholesteric phases 84
– of smectic C* phases 89
unwinding of the helix
– of cholesteric phases 84
– of smectic C* phases 89

V
Vectra 116
viscosity 99
– coefficients 100
– rotational 100
visible spectroscopy 180
– nematic solvents for 175,
  176
Vuks equation 58

W
Williams-Kapustin
  domains 95

X
X-ray diffractograms 13, 15
X-ray studies in lyotropic
  mesophases 160
X-ray studies in polymeric
  LC 129
X-ray studies of thermotropic
  mesophases 13
Xydar 116

MIX
Papier aus verantwortungsvollen Quellen
Paper from responsible sources
FSC® C105338

If you have any concerns about our products,
you can contact us on
ProductSafety@springernature.com

In case Publisher is established outside the EU,
the EU authorized representative is:
**Springer Nature Customer Service Center GmbH
Europaplatz 3, 69115 Heidelberg, Germany**

Printed by Libri Plureos GmbH
in Hamburg, Germany